Reason Awake

Science for Man

Institute for the Study of Science in Human Affairs of Columbia University

Reason Awake: Science for Man is sponsored by the Institute for the Study of Science in Human Affairs. It had its origins in a series of lectures delivered under Institute auspices in observance of the two hundredth anniversary of the College of Physicians and Surgeons and with the support of the Josiah Macy, Jr., Foundation.

The Institute was established in 1966 as a permanent part of Columbia University and in that year received a founding grant from the Alfred P. Sloan Foundation. Its task is to study science and consequent technologies in the context of human affairs: how science affects society and society science and science institutions. It seeks, through research inquiries, interdisciplinary projects, curriculum developments, and publications, to advance basic understanding of the human implications of new knowledge in the physicial and biological sciences and of the conditions for more effective scientific research and development, more knowledgeable and appropriate application of the resulting discoveries, and more meaningful general understanding of science in human affairs. In addition to sponsoring books which emanate from Institute activities, the Institute publishes series of monographs, bulletins, and occasional papers.

Reason Awake

Science for Man

René Dubos

Columbia University Press
New York and London 1970

René Dubos is Professor of Environmental Biomedicine at The Rockefeller University.

SBN: 231-03181-5
Library of Congress Catalog Card Number: 70-111327
Printed in the United States of America

This book is dedicated to "all men who think that the future belongeth unto them."

Louis Le Roy
(1594)

Foreword

At the time Columbus set sail across the Atlantic the coat of arms of the Spanish royal family was an *impressa* depicting the Pillars of Hercules, the Straights of Gibraltar, to symbolize that Spain was at the outpost of the world. Its proud motto was *Ne Plus Ultra,* "no more beyond." After the discovery of America the Pillars of Hercules remained on the coat of arms, but the motto was changed to *Plus Ultra.* There was "more beyond," and Spain regarded herself as the gateway to the new world.

The motto *Plus Ultra* became the rallying cry of Francis Bacon's followers, who identified the future with experimental science, and it has continued to express the euphoric attitude of the scientific community. Vannevar Bush was the spokesman for his generation when he wrote in 1945 that science, the endless frontier, now constitutes the great challenge for the spirit of man and for his energies.

Recently, however, doubts have been expressed concerning the possibility and even the wisdom of exploring beyond what is already known. In his latest book, aptly entitled *Science Is Not Enough* (1967), Vannevar Bush stated that there are limits to human understanding: "Science proves nothing absolutely. On the most vital questions, it does not even produce evidence." Furthermore, men in all social classes, and scientists in particular, are beginning to fear that the applications of science may threaten the survival of man and certainly despoil the quality of life.

I shall quote in several chapters of this book stern warnings to this effect by leaders of American science. But I could have selected statements just as forceful from European scientists—for example, these words by Professor Egan Orowan, a British physicist who is a Fellow of the Royal Society: "The vast majority of the Earth's population regards science and technology as an increasingly mortal threat to their lives. They feel themselves powerless at the mercy of a few, as if they were on the operating table in the hands, not of healers, but of irresponsible playboys driven by curiosity." Further, according to Orowan, scientists should realize "that they are dancing on a powder keg." Oddly enough, one of the most extreme, and in my opinion unjustified, expressions of concern about the wisdom of developing all aspects of scientific knowledge has come from a scientist-philosopher who has repeatedly and ably pleaded the cause of scientific technology. According to Mesthene (1967), we may soon have "to reconsider the wisdom of the traditional belief in the 'duty' of science to explore the unknown unhampered by any other considerations."

A brief statement of some of my own beliefs may help the reader evaluate what I shall have to say throughout this book concerning man's attitude toward science and the role played by scientific knowledge in human life and in the development of civilization. In my judgment social constraints on the scientific enterprise have been made inevitable by the fact that science impinges with increasing effectiveness and violence on all aspects of human life. Scientific technologies affect the well-being of man and the ecologic balance of the planet; before long the discoveries of the behavioral and social sciences will make possible the manipulation of individual human beings and entire populations. Consequently, the need to evaluate the full implications of science in human affairs will call for increasing concern with the direction of the scientific effort and with the comparative rates at which the various scientific disciplines should be developed.

Ralph Waldo Emerson was responsible for the phrase, "Things are in the saddle and ride mankind," but before Emerson the same thought had been expressed several times in the course of human history. What neither Emerson nor his predecessors could have realized, however, is the extent to which our age would be obsessed with "things"—their manufacture, their distribution, their use. The word *progress* has been corrupted because it is now used almost exclusively to denote changes in the "standard of living," an expression that refers not to the quality of life, but to the abundance of things. Progress no longer means a higher degree of education, more enlightened tastes, nor even better health; rather, it has come to denote how many manufactured objects people can own, how many destructive weapons a nation possesses, or, at best, how many space vehicles a nation can put into orbit or deposit on celestial bodies. As the power of science increases, its uses become less sacred, more trivial, more brutal, and often more immoral. Scientists are not entirely responsible for this desecration of their achievements, but we have done little to prevent it. As a community we have betrayed our ideals by vaunting our trades and promoting our wares through irresponsible promises to society of perfect health, economic prosperity, and military power. We may even degrade the intellectual quality of our work, if we praise exclusively the advancement of science and its competitive aspects and fail to integrate scientific knowledge in civilized life.

Much lip service is still being paid to the notion that science is primarily concerned with pure knowledge and has little if anything to do with the practical affairs of human life. But in fact most of scientific knowledge is now concerned with practical problems. It seems to me unwise and ambiguous for scientists to affirm on the one hand that they are primarily searchers for truth and to claim on the other hand that everything they do is ultimately of practical importance. This ambiguity creates in the general public the feeling that scientists engage in double talk to provide a rationalization for what they really want to do, under the pretext that their findings will eventually be of social use. What is often called the anti-scientific movement is probably little more than an expression of the fact that the public is losing confidence in the ill-conceived and usually exaggerated claims of the scientific community.

Important as they are, the technological and other practical applications of science have been oversold. In fact, the production of goods and the development of what is now called technological "fixes" may not be the most valuable contribution that scientists can make to society. Of probably greater usefulness would be the development of knowledge and attitudes that would help man to examine objectively, rationally, and creatively the problems that are emerging as a result of social evolution. But this aspect of science is given very low priority—if not neglected altogether—in universities and research institutes. Academics and technologists proudly proclaim that we live in an age of science. What this really means is that we exploit the world's natural resources, usually without regard to genuine human needs, and we can correct a few disorders of the body and the mind without concern for happiness. We hardly give any thought to the long-range consequences of our scientific and technological interventions into man's life and nature.

I shall now open a large parenthesis to outline a few thoughts I have formulated about the near future, after *Reason Awake* was in proof, too late for incorporation in the main text.

Like many other scientists and laymen, I have become convinced that industrial societies will inevitably meet disaster if they continue producing more and more of everything, for larger and larger numbers of persons, as they have been doing at an ever-increasing rate during the past century. Here are a few of the reasons, dogmatically stated, which lead me to believe that the quantitative expansion of technology will soon come to an end.

(a) The world population will stop growing and may even fall under the pressure of different forces. In some parts of the world this will happen as a result of food shortages; biological disasters such as mass disease and mass poisoning are likely to occur in other areas; willful control of birthrates will be achieved in a few countries.

(b) The amount of energy used for industrial and domestic purposes will eventually reach a plateau—even if new kinds of low-cost fuels become available and if the production of "clean" nuclear energy becomes technologically possible. The limitation will come not from shortage of energy sources, but from the fact that the injection of excessive amounts of energy into natural systems inevitably disturbs their operations and commonly leads to ecological disasters.

(c) The quantity of things produced by technology will also reach a plateau, because of shortages in certain natural resources and because environmental pollution will reach unbearable levels. The present accumulation of solid wastes—chemical and organic—is a portent of worse things to come.

All ecological systems, whether man-made or natural, must in the long run achieve a state of equilibrium and be self-regenerating with regard to both energy and materials. The ecology of highly industrialized nations has been in a state of disequilibrium for several decades. Furthermore, ecological instability is increasing at such an accelerated rate that disasters are inevitable if the trend continues. We cannot afford to delay much longer the development of a nearly "closed" system in which materials will retain their value throughout the system by being recycled instead of discarded.

The ecological constraints on population and technological growth will inevitably lead to social and economic systems different from the ones in which we live today. In order to survive, mankind will have to develop what might be called a "steady state"—a phrase that I prefer to John Stuart Mill's "stationary state."

The "steady state" formula is so different from the philosophy of endless quantitative growth, which has governed Western civilization during the nineteenth and twentieth centuries, that it may cause public alarm. Many persons will mistakenly assume that the world is entering a period of stagnation, leading eventually to decadence. Yet, a steady state is compatible with creative changes. In fact, change within a closed system will probably offer intellectual possibilities much more challenging than those offered by the kind of rampant growth that prevails at present.

The ecological constraints on the growth of the world population and on the production of energy and of goods will generate new kinds of scientific problems. For example:

—the drastic limitation of family size will probably create social, psychological, physiological, and perhaps even genetic disturbances concerning which little, if anything, is known.

—the distribution and utilization of energy under controlled conditions will require sophisticated knowledge of regional and spaceship ecology.

—entirely new technologies, and therefore new kinds of scientific knowledge, will have to be developed to minimize pollution and to recycle the natural resources in short supply.

The steady state will thus compel a reorientation of the scientific and technologic enterprise. Indeed, it may generate a scientific renaissance. But this will not happen without a conscious, and probably painful, effort from the scientific community.

So far, universities and research institutes have largely remained aloof from the problems that the world will face in an acute form before this century is over. The pressure of public opinion, however, will soon force scientists out of this aloofness. Scientists will have to redirect their thoughts and skill away from the problems in which they are now interested, toward problems of larger social significance. Rapid and profound shifts in areas of emphasis will therefore occur with regard to theoretical science and to technology.

Foreword

New scientific concepts emerge from science itself, either as products of its own internal logic or through accidental discoveries which present some analogies to the mutations of the biological world. This aspect of the advancement of knowledge might be called the internal history of science. Equally important is the external history of science, because the development of a new concept, and especially its conversion into a form which is meaningful for society at large, is profoundly influenced by the social milieu. In science as in other human occupations, the success of an idea depends not only on its own structure and merit, but also on the nature of the ground upon which it falls, and on its suitability to social needs. The constraints inherent in the world of the immediate future do not imply a retreat from science. But they make ideas concerned with the design of the scientific enterprise, rather than accumulation of facts related to growth, the dominant need for the advancement of science and of technology.

As will be evident from the tone of this foreword, the central theme of the present book is not science per se but rather the penetration of science into all aspects of human life. It is in this light that I shall discuss the role and the rights of scientists in the social order. When considered as a purely intellectural pursuit, science is the private affair of scientists, just as literature and the arts are the private affairs of artists and writers. But science becomes a matter of public concern whenever it is applied to social problems. In practice most scientists have now become public servants because they assert that their work contributes to mankind's welfare and therefore deserves greater financial support than that given to the humanities. If scientists expect a greater share of the public resources than other human activities, they must justify their claims by the ensuing benefits their work brings to society. And they must consequently allow society to be the final judge of the kinds of science that are most worthy of support.

The lectures from which this book was prepared were delivered at the invitation of the Institute for the Study of Science in Human Affairs of Columbia University. Hence, their emphasis on utilitarian considerations and the urgent need to correct the social absurdities and monstrosities that result from the mismanagement of scientific technology.

I wish I had had the opportunity to discuss also the pursuit of science for its intrinsic value without regard to profitability, because this aspect of scientific endeavor may eventually become one of the most interesting expressions of human life. The acquisition, organization, and assimilation of scientific knowledge constitutes a major intellectual and artistic enterprise, together with philosophy, music, writing, painting, or acting. There is an aspect of science that transcends pedestrian utility yet has immense social relevance because it enhances humanness. It is "the privilege of man to understand," wrote Vannevar Bush (1967). For this reason, "It is the duty so to live that there may be a reason for living, beyond the mere mechanisms of life. It is the duty to carry on, under stress, the search for understanding."

These words call to mind the passage in Dante's *Divine Comedy* in which Ulysses exhorts his companions to go on: "You have your lives, not so that you may live like beasts, but rather that you may strive for fame and knowledge." In Tennyson's poem *Ulysses*, the hero, also urges his companions "to sail beyond the sunset":

> To follow knowledge like a sinking star,
> Beyond the utmost bound of human thought . . . .
>
> Death closes all; but something ere the end,
> Some work of noble note, may yet be done.

In the final analysis the greatest social contribution of science may well be to help man shape his own destiny by giving him knowledge of the cosmos and of his own nature. Through science we can learn about the world around us, how we emerged from it, what we can do with impunity, and the best way to reach our goals. We can even learn to formulate new goals compatible with our fundamental nature and with the constraints imposed on us by the natural forces of which we are the expression. When man truly enters the age of science he will abandon his crude and destructive attempts to conquer nature. He will instead learn to insert himself into the environment in such a manner that his ways of life and technologies make him once more at harmony with nature.

René Dubos

Acknowledgments

I am indebted to a number of publishers and individuals for permission to quote from their works. They include *American Scientist;* American Society for Public Administration; *Bulletin of Atomic Scientists;* Columbia University Press; *Daedalus; Encounter;* Little, Brown and Company; M.I.T. Press; *Massachusetts Institute of Technology Review; Science;* Dr. Frederick Seitz; *Stanford Alumni Almanac;* American Institute of Physics; and University of Washington Press.

The bibliographical details for all quotations are given in the text and in the Selected Bibliography.

R. D.

Contents

1
The Despairing Optimist

The remark made to Hannibal by one of his officers at the end of the second Punic War speaks to Western civilization, especially in the United States: "You know how to win victories, Hannibal, but you do not know how to use them."

Hannibal's bold generalship and his imaginative cavalry tactics, including the use of elephants, enabled him to defeat one Roman army after another; but he ultimately failed because the social institutions of Carthage did not give him adequate support and, more importantly did not formulate worthwhile goals for his military genius. Similarly, Western civilization has displayed immense ingenuity and vigor in using scientific technology for the exploitation of nature; nevertheless modern societies are deteriorating because they are inept in applying their scientific and technological prowess to the concerns of human life.

The Despairing Optimist

Everywhere in the Western world the amenities of existence are threatened by ecological degradation and existential nausea. The mounting roster of environmental and psychological problems creates the impression that mankind has lost control of its affairs. The size and form of our cities, our contacts with nature, the occupations in which we engage, and the very directions of our lives are determined more by technological imperatives than by the choice of desirable human goals.

Science and technology, which are creations of man, cannot be responsible for his ineptness in using knowledge and power. But we scientists and technologists have promised so much more than we can deliver that scientific technology has now become the scapegoat for civilization's failures. A large number of human beings, especially among educated young adults, would agree with what Anatole France wrote half a century ago: "I despise science for having loved it too much, like those disenchanted voluptuaries who reproach women for not having given them the satisfactions they had expected from love." Max Weber was prophetic when he asserted, early in this century, that disenchantment would be the mark of our times.

The disenchantment with civilization is now so widespread that it pervades the daily press. In the *New York Times* (1967) James Reston introduced the phrase "new pessimism" (which was reprinted in an editorial of the *Wall Street Journal*, 1967) to denote the attitude of those who believe that many problems of the modern world are caused by scientific technology yet are not amenable to scientific control.

Reston was perceptive in using the word pessimism to describe the present attitude of the educated public toward science and technology, but this attitude of pessimism is not a new one. The social consequences of increased knowledge and technological innovations caused as much uneasiness and alarm among Western Europeans 400 years ago as they do among us today.

In 1575 the French scholar and jurist Louis Le Roy—so learned and wise a man that his contemporaries referred to him as the French Plato—published a book on the cultural history of mankind. He gave it the ambitious and puzzling title *De la Vicissitude ou variété des choses en l'univers, et concurrence des armes et des lettres par les premières et plus illustres nations du monde, depuis le temps ou a commencé la civilitè, & mémoire humaine iusques à present*.

Le Roy made it a point to discuss in detail the history of technological inventions from the earliest developments, because he wanted to show that each historical period has added something to man's understanding and control of nature. His main purpose in writing the book, however, was to discuss the effects the new learning and new inventions were having on human life in the sixteenth century and would have in the future.

Le Roy's *Vicissitude* is rather dull reading for us, but the book must have struck a sensitive nerve when it appeared in Europe, as judged from the fact that it was reprinted in Paris in 1576, 1577, 1579, 1583, and 1584. It was translated into Italian in 1584, reprinted in 1592, and finally was translated into English in 1594 under the title *Of the Interchangeable Course or Variety of Things in the Whole World.*

According to V. Harris in *All Coherence Gone* (1949), Le Roy's book was widely used in England during the seventeenth century as a source of arguments among humanists who discussed the decay of nature. There is also evidence that it influenced Francis Bacon's view concerning the social implications of science.

When Le Roy wrote his *Vicissitude* the Renaissance was well under way. It had brought about profound changes in religious beliefs as well as in many intellectual and social aspects of European life. The changes gave reason for both hope and despair.

The invention of the mariner's compass had made it possible to discover lands beyond the oceans unknown to the ancients, and it had brought distant regions near. Le Roy's enthusiastic discussion of this topic indicates how much he and his contemporaries were exhilarated by the feats of the great navigators and by the geographical discoveries of his time. Exploration, however, had exposed man to new diseases. Syphilis, perhaps introduced into Europe from the West Indies by Columbus' crew, had spread like a conflagration. The prevalence of venereal infection was increased by both the general relaxation of sexual mores and the mass movements of armies and populations at the beginning of the sixteenth century.

The Despairing Optimist

The printing press had made new information and ideas available to an extent and at a rate undreamed of at any time before. Le Roy marveled at the immense gains in knowledge he had witnessed, and was particularly impressed by the creation of public libraries that greatly facilitated the dissemination of knowledge:

> The princes who have helped most to restore the arts are Pope Nicholas V, and Alfonso, King of Naples, who received with honor and liberally remunerated those who presented to them translations of Greek and Latin books. The King of France, Francis I, endowed public professorships at Paris, and prepared a lavish library, full of good books, at Fontainebleau. Without the favor and liberality of the kings of Castile and Portugal, no one would have come to discover the new lands or sail to the Indies. The Florentine lords of the house of Medici, Cosimo and Lorenzo, helped very much, receiving learned men who came to them from all over, and these they maintained honorably. At their own expense, they sent people to search throughout Greece for the good and ancient books that were lost. They built magnificent libraries for the common use.

The easier communication of knowledge, however, had generated confusion in the public and precipitated intellectual crises among the bewildered scholars. The ancient loyalties which had so far bound the social order were breaking down under the impact of the new enlightenment, the intellectual freedom, and especially the wars of religion. Le Roy was appalled by the extent of the social unrest and the menace of universal war. It seemed to him that an era of darkness and perhaps self-destruction might be at hand.

To make things worse, the introduction of gunpowder and the improvements in firearms had made all ancient weapons obsolete. The art of war was being transformed into something much more destructive and deadly.

Wherever one looked, north or south, east or west, there were social disturbances, intellectual uncertainties, religious wars, and political upheavals. In a display of economic awareness rare among the humanists of his time, Le Roy even noted the rise in prices that was occurring throughout Europe during the sixteenth century!

Despairing of his times, he exclaimed, ". . . for a long time there has not been more malice in the world, more impiety, and disloyalty. Devotion is extinct, simplicity and innocence are mocked, there remains only the shadow of justice. All is pell-mell, confounded, nothing goes as it should."

Despite all these upheavals and new threats Le Roy tried to remain optimistic about the future. He believed in particular that the advances in transporation and communication would make it possible for "all mortals to exchange their goods among themselves and help each other as inhabitants of one community and one world commonwealth." Wendell Wilkie's "One World" and Adlai Stevenson's "Spaceship Earth" appeared even then as possible consequences of advances in knowledge!

The Despairing Optimist

History never repeats itself and should not be used uncritically for the interpretation of the present or as a guide to the future. I have nevertheless found it justifiable to review Le Roy's *Vicissitude* at some length because his book illustrates that many of the social problems commonly assumed to have emerged as a result of the impact of modern science and technology on human affairs actually have historical precedents. Upheavals have occurred in the past wherever and whenever the ways of life have been upset by social or technological innovations or sudden spurts in the growth of general knowledge. It is historically naive, and perhaps an expression of collective conceit, to assume that we are faced with entirely new problems. Modern science is not the only force that has qualitatively altered—for good or for evil—the conditions of man's life and the quality of his relationships with other human beings and the rest of creation. From Christianity to Marxism, from the improvement of sailing in the fifteenth century to the introduction of the railroad in the nineteenth century, many are the forces that have turned the world upside down without benefit of theoretical science.

Le Roy's *Vicissitude* is relevant to the present book for another reason, namely because it provides a pattern for the attitude I shall take in discussing the scientific enterprise and its technological applications. I do not intend to discuss science as a method of inquiry, a body of knowledge, or a philosophical view of creation. As was Le Roy's concern, mine shall be only with the interplay between science and human affairs—the undesirable as well as the beneficial effects.

The fact that I have used the words "undesirable" and "beneficial" in the preceding sentence implies that I shall introduce value judgments in my discussion. All knowledge is good, but not all advances in knowledge are equally good or urgent. Also many applications of knowledge that are possible are not desirable, and some are objectionable. Scientists and technologists are not responsible for the use society makes of their achievements, but we are guilty of escapism and irresponsibility if we do not concern ourselves with the social consequences of our work. Since the public supports us and puts its trust in us we must publicize the potential dangers of scientific advances as well as their obvious benefits.

Hope and pessimism will compete in my evaluation of the scientific enterprise. If hope predominates it is because I believe that optimism is essential for action and constitutes the only attitude compatible with sanity. Hope, however, will always be qualified by doubt. As did Le Roy, I shall consider the present form of scientific civilization in the mood of a despairing optimist.

I despair at seeing experimental science, one of the sanest and most powerful instruments ever devised by the human mind, used for purposes which are the epitome of human folly. Ecological disasters and decay of the spirit are inevitable if we let scientific technology continue on its present course. Many societies in the past have committed themselves to a course that eventually led to their destruction. Similarly, ours will experience catastrophes if it remains obsessed as it is today with the production of more power and more things.

On the other hand, I am an optimist because the word catastrophe has not always implied tragic upheavals and utter failure. The original Greek word *catastrophe* meant a sudden change of course, an overturn not necessarily associated with disaster. There is still time for Western civilization to change its course by refocusing scientific technology on goals that are socially desirable and compatible with human biology. One of the most hopeful signs of our period is the extent of soul-searching among scientists and sociologists concerning the nature of the scientific enterprise and the urgency to rededicate it to worthwhile social goals.

The philosophers of the Enlightenment had learned from Bacon that knowledge is power, but they also believed that power should be used for the betterment of mankind. In the United States Jefferson and Franklin were typical representatives of the belief that scientific knowledge would eventually enable man to shape the physical and social environment according to his wishes. In Europe one of the most explicit expressions of this faith was the *Sketch for a Historical Picture of the Progress of the Human Mind*, written by the Marquis de Condorcet in 1793 amidst the turmoil of the French Revolution. Advances in science and technology, improvements in nutrition, medicine, and housing, social equality and political independence for all, and equal rights for women were all envisioned by Condorcet and his contemporaries as inevitable future consequences of scientific knowledge.

Were they to come back to life the philosophers of the Enlightenment would be startled to find that it took less than 200 years for many of their forecasts and hopes to become realities, at least in Western civilization.

Although we have become somewhat blasé about the marvels of our age, their magnitude can be recaptured by trying to imagine what our existence would be without them. Medicine has come close to solving many problems of disease that made human life precarious as late as the nineteenth century. Nutritional science has determined the essential food requirements of man, and technology has made it possible to meet these requirements at all seasons in any climate. Almost everyone in the Western world can stay warm during the winter and soon will be able to cool his dwelling during the summer. The barriers of distance become less of a problem with each passing day, and neither lack of time nor strength need any longer limit our ability to move anywhere over the globe. From penicillin to the control of personality, from the invention of synthetic fibers to space exploration, the twentieth century has been marked by achievements so startling that they dwarf the miracles of legendary ages.

Furthermore, the revolutionary advances of the past two centuries warrant the conviction that almost any problem can be solved by scientific knowledge if it is properly formulated and its study diligently pursued. As a student of experimental medicine I am convinced that progress can be made in the control of any disease to which we address ourselves with energy and imagination. Similarly, I feel confident that physicists, chemists, and engineers can provide us with almost any kind of earthly good. Most enlightened persons now take for granted that scientific research will increasingly enable man to master natural forces and control his own nature. Yet, despite the achievements of the immediate past and the general belief that many more advances are in the offing, the future appears uncertain and even frightening. As in Le Roy's time, today's revolution in technology and knowledge creates an age of anxiety.

The most obvious reason for disenchantment is the realization that prosperity and comfort do not assure health and happiness. In fact, material progress often has consequences that spoil the quality of life. Environmental pollution, the crescendo of sensory stimuli, especially noise levels, the progressive erosion of public services, the loss of privacy, and the increase in social regimentation are but a few of the distressing aspects of modern life that originate from technology, or at least from its mismanagement.

Another cause of anxiety is the fact that scientific knowledge has weakened or destroyed the traditional values by which man functioned in the past, but it has not provided him with a new ethical system to serve as a substitute. Knowledge has freed man of a few gross errors and paralyzing fears, but it has not given him beliefs that would add a joyful spirit to material existence. Science, it has been said, gives man everything to live with but nothing to live for.

Many world problems are of course social rather than scientific or technological in origin. If the population continues to grow beyond the capacity of the earth to support it adequately and to absorb its waste products, and if man expands still further his appetite for the creations of industry, no control measure can possibly cope with human excesses. In the final analysis civilization can be saved only if we are willing to change our ways of life. We have to invent utopias, not necessarily to make them reality but to help us formulate worthwhile human goals.

I shall not attempt the formidable task of imagining utopias for our times, but I shall try to illustrate with a few examples how scientific research could be redirected to problems far more important for human welfare than those which are receiving much attention at the present time. I shall take specific examples from the program of the scientific meetings held in December 1967 to celebrate the Sesquicentennial Anniversary of the New York Academy of Sciences. This Academy is of special social interest because it has been intensely concerned, throughout its existence, with the applications of science to all aspects of human life.

Of course nuclear science occupied a privileged place on the anniversary program, and confidence was expressed that mankind would soon be provided with endless sources of power. Experience has shown, how-over, that almost any form of power, if used without regard to future consequences, degrades the environment and decreases its fitness for human life. A more sophisticated science of ecology could provide knowledge of the kinds of social and technological manipulations that are compatible with a healthy environment, but this knowledge is undeveloped because of lack of support in universities and research institutes. Indeed, ecology was not represented on the anniversary program of the New York Academy of Sciences. Yet nuclear technologies, or any other technologies, should not be developed in the vague belief that they can be used indiscriminately whenever convenient and economically profitable. All technological innovations should be tailored to fit social goals based on ecological considerations.

Chemists and engineers will unquestionably produce many new kinds of materials and processes that can change human life. It is commonly assumed that man must adapt to these changes but in fact human adaptability is not limitless. We know little of the thresholds and ranges of human adaptability. In any case it is certain that the acceptance of social and technological changes does not mean that these are necessarily desirable. It is widely recognized that ionizing radiations and environmental pollution of air, food, and water have deleterious effects that manifest themselves very slowly; they behave like the pestilence that stealeth in the darkness. Similarly, many social and technological innovations that appear readily tolerated eventually ruin the quality of human life. The actual limits of adaptability are not determined by what can be tolerated for a limited period of time, but by future consequences. These consequences are essential factors to be considered in deciding what innovations are safe and desirable.

One of the papers presented on the occasion of the Sesquicentennial Anniversary dealt with heart–assist devices. This presentation was very timely because heart surgery and transplantation are now being spotlighted by the medical community and the press. There is no doubt that progress will be made in the further development of artificial organs and in the surgical and immunological aspects of organ transplantation. But it would be equally interesting scientifically and of much greater social usefulness to learn how to govern life so as to make such devices and surgical procedures unnecessary. Under the proper conditions man's heart can probably function effectively for at least 100 years. Search for the factors in our present ways of life and environment that are responsible for the increase in vascular disorders in general and heart disease in particular should have high priority in medical research, if only because prevention is always far more effective than cure, as well as being much less costly and applicable to many more persons.

Another paper at the Sesquicentennial celebration of the New York Academy of Sciences dealt with communications and mass media. This was also a timely topic since global man could theoretically now replace parochial man. Technological processes enable him to read, hear, and see almost anything that goes on in the world. In practice, however, technology is only a small part of the problem of communication. A breakdown of intellectual contacts is occurring everywhere. If we take the scientific community as a typical example, it is obvious that scientists communicate poorly, if at all, with the public and that they even have great difficulty communicating among themselves. The specialist discusses matters relevant to his profession almost exclusively with a few other specialists of his guild; and the same narrowness of intellectual intercourse holds true in areas of knowledge outside of science. Mass media cannot possibly be of any help in improving this fundamental deficiency in communication.

We as members of specialized scientific communities develop concepts and jargons which only we can understand, resulting in a lamentable impoverishment of the public sector of our lives. Yet it is probable that most of us could learn to recast our professional knowledge in terms meaningful to nonspecialists. The important problem therefore is not one of developing more rapid and precise mechanisms of mass communication, but rather learning how man (scientist or not) can speak to man directly about experiences he can share with many human beings.

Knowledge of the natural sciences is being applied to a great variety of practical problems. Many of the applications, however, are undesirable, or at least much less urgent that others which are grossly neglected. One of the shockingly neglected fields of science concerns the effect of environmental factors on the development of human characteristics. All scientists and other academics proclaim that culture is the ultimate good and that all citizens must be given the same educational opportunities. However, this cannot be done effectively until we have acquired more knowledge concerning the development of mental potentialities and the receptivity of the human organism to various kinds of stimuli and information. We should make a vigorous effort to understand the effects that prenatal and postnatal influences exert on biological and mental characteristics and to determine the responses of the human organism to environmental stimuli at various critical periods of life.

Eventually scientists will have to concern themselves with the problem of human happiness, but this is probably asking too much too soon. We should at least investigate more diligently some of the factors that bear on happiness for example: the relationships between physiological characteristics, behavioral patterns, social structures, and the formulation of attainable goals in desirable environments.

Most needed, perhaps, is a better knowledge of the range of man's physical and mental potentialities. The word "potentialities" must be used in the plural because there is overwhelming evidence that under normal conditions the ways of life and the surroundings do not allow the full expression of the capabilities inherent in man's genetic constitution.

When the signers of the Declaration of Independence proclaimed that the pursuit of happiness was an inalienable right of man, they emphasized not happiness per se, but the freedom of every man to pursue happiness by ways of his own choice. Science is socially successful to the extent that it increases the range of freedom and the number of options available for this pursuit. Its continued support by society will depend on the willingness of scientists to relate their professional interests to genuine human needs and goals. Science is not only an exciting adventure for scholars and a technique for economic expansion; it is also an essential part of the social organism and must continuously evolve under the influence of changing social conditions.

2

Natural Philosophers, Inventors, and Scientists

Words have clearly defined dictionary meanings which give the illusion of exactness. In the spoken or written language, however, these meanings are enriched with overtones derived from the history, beliefs, social attitudes, and professional occupations of the speaker or writer. As pointed out by Benjamin Franklin, "Such is the imperfection of our language, and perhaps of all other languages, that, notwithstanding we are furnished with dictionaries innumerable, we cannot precisely know the import of words, unless we know of what party the man is that uses them."

Consider, for example, what the word science means when applied to the intellectual activities represented in Raphael's fresco *The School of Athens*, now in the Vatican Museum. Plato and Aristotle occupy the center, surrounded by Pythagoras, Heraclitus, Diogenes, Archimedes, and other illustrious representatives of Hellenic culture, each in a characteristic pose, carrying the tools of his specialty. We may have forgotten the specific achievements of the scholars in Raphael's fresco, but we know that they have provided the intellectual basis for what we call modern science. Hellenic culture gave us faith in the belief that rational thinking and verifiable facts constitute the stuff of science and that scientific knowledge helps man elevate his life above brutish existence.

The scholars represented in *The School of Athens* were concerned with understanding the world rather than with increasing man's wealth or physical comforts. They represent knowledge that was once called natural philosophy, rather than skills that have produced technology. Even in the ancient world, however, science could take forms very different from those which prevailed in classical Greece.

Joseph Needham (1954) has described in fascinating detail the prodigious achievements of ancient Chinese technology based on empirical knowledge and skills acquired long before the medieval era. According to Needham Chinese science derived its empirical emphasis from the study of water conservation and the techniques of irrigation; these technological problems were of such crucial importance in ancient China that their study and administration came very early under governmental control. Throughout China the invention of mechanical contrivances and the acquisition of knowledge of natural phenomena spread far beyond the technology of water. For example it extended to such an esoteric chemical and biological topic as the separation of sex hormones from the urine of pregnant animals! The theoretical knowledge and practical achievements of Chinese scholars and inventors led to conclusions and generalizations that are still valid today. Yet, this knowledge was obtained by trial and error rather than developed from the kind of rational analysis which was such a striking feature of Hellenic civilization.

As presently used the word "science" thus has several different meanings, each one of them legitimate. It denotes the formulation of the laws of nature and the description of substances, events, and behaviors in terms sufficiently broad and abstract to encompass as large a set of situations as possible. On the other hand, it also applies to the development of particular products and techniques—penicillin and color television, pasteurized baby foods and atomic reactors—in brief, all the things and procedures, either desirable or objectionable, that are made possible by technology.

If there is one common denominator to these two different kinds of science, it is the verifiability of the assertions concerning the laws of nature, the observed phenomena, or the practical application under study. This verifiability makes science the public knowledge par excellence (Ziman, 1968); but knowledge is valid only within a certain range of conditions. The scientist must be satisfied with present verification without ultimate certainty (Calder, 1968).

Philosophers, Inventors, Scientists

According to Arnold J. Toynbee (1968),

> There have been many definitions of the word "science." Perhaps the most generally accepted one is that science is a form of study in which there can be an exact knowledge of the present and the past and, through this, an infallible prediction of the future. If this is what science means, then no study made by a human mind can be completely scientific. But there will be differences in the degree of approximation to scientific study, and these differences will be determined by the nature of the part or aspect of the Universe under consideration. Study will be most scientific when its object is the physical structure of the Universe. . . . The object of study that will be the least amenable to scientific treatment is the non-physical facet of human nature. Students in this field had better avoid letting themselves be tempted by the present-day prestige of the word "science" into applying that label to their own work.

Few, if any, scientists would approve of Toynbee's definition or accept the limitations that he wishes to impose on the fields to be studied by the methods of science. Progressively, the scientific approach is being extended to all manifestations of nature, including man, about which verifiable knowledge can be obtained. It is even being applied to such elusive phenomena as the spontaneous emotional interplay between mother and child as well as to the forces that impose a pattern on the size of animal populations and on business cycles. Certain types of social and behavioral studies can unquestionably achieve a degree of verifiability and predictability sufficient to be considered scientific, especially when they deal with large numbers of living organisms, animal or human.

The vagueness of the word "science" has its explanation in a host of historical factors. Marxist historians have defended the view that all of science derives from attempts to solve practical problems, whereas, around the turn of the century most scientists believed that science was pursued for its own sake and that its chief purpose was to discover the laws of nature. In reality, we do not even know whether ancient man observed the celestial bodies because he wanted to use them as guides for navigation or because he was curious about their nature and their motions. Similarly, we do not know whether ancient man examined, described, and classified animals and plants because he used them as a source of food and drugs or because he was perplexed by their very existence or by the nature of life. In my opinion the pursuit of practical ends and the acquisition of knowledge have been inseparable aspects of the life of *Homo sapiens* from its beginnings. In fact, becoming scientific and acquiring humanness probably occurred simultaneously in the prehistory of mankind.

The words "knowledge" and "science" acquire different values according to the attitude toward the future taken by the social groups which use them. Arcadian types of cultures look to the distant past as the time of perfect bliss and therefore regard the mores and customs inherited from ancestors as the most valuable kind of knowledge. They try to preserve their ways of life and beliefs and transmit them unchanged from one generation to the next. Such traditional knowledge is always rich and often factual, incorporating as it does accumulated group experiences that have been tested under the practical conditions of daily life. This kind of knowledge, largely derived from observations rather than from experimentation, is very different from what modern man calls scientific knowledge because the latter is constantly being transformed by the experimental method.

In the utopian types of cultures man believes that he will achieve happiness only after he has built the new Jerusalem. Utopias differ in character one from another because they reflect the tastes and value judgments of their originators. Some utopias take anarchism as their ideal; others try to achieve Nirvana; still others long for the passivity and tolerance of Goncharov's oblomovism. The characteristic ideal of Western civilization, however, is the ceaseless activity and creativity of the Faustian universe. Faustian man values above all those forms of knowledge that enable him to master natural forces and transform the world. In the countries of Western civilization science means chiefly experimental science because experimentation suits the searching mentality and restless temperament of Faustian man.

Most human beings are now ambivalent in their attitude toward the past and the future. As life becomes more complex and man is increasingly deprived of fundamental satisfactions, he longs for simpler ways of life and daydreams about the happiness associated with traditions and legends in less developed societies. Nevertheless, the belief in Arcadia is progressively vanishing all over the world as evidence is mounting that "the world we have lost" was not as simple and idyllic as we used to imagine (Laslett, 1965). In any case, most human beings now behave as if there were no alternative to the dynamism of Faustian life, with the result that the concept of culture is changing accordingly. Whereas traditional civilizations had put a premium on the *preservation* and faithful transfer of customs and skills, modern societies tend to consider that the most serious concern of culture is the *advancement* of knowledge, preferably through the techniques of experimental science.

The people of ancient China, India, Greece, Rome, Islam, and pre-Columbian America certainly had the intellectual capacity for developing experimental science. They could not have accumulated and organized their objective knowledge without experimentation. Yet they never pursued experimental science on a large scale. Conversely, since the seventeenth century people of European origin have been obsessed with the thought that knowledge should be advanced by experimentation as rapidly as possible; and they have believed that the experimental method would enable them to improve the world and to increase happiness. Increasingly they have become more interested in the advancement of knowledge than in its possession and integration.

It is not known why the experimental method merged with such vigor in seventeenth century

Europe, but there is no doubt that Francis Bacon was one of its first and most articulate exponents. He based the new scientific faith on the Christian dogma of Original Sin and on the belief that man could recover his dominion over the natural world by a systematic cultivation of knowledge. According to Bacon the inductive method was a foolproof way to achieve this end. In his words, "Whereas in the past the proceeding has been to fly at once from the senses and particulars, up to the most general propositions . . . my plan is to proceed regularly and gradually from one axiom to another, so that the most general are not reached till the last; but then when you do come to them you find them—such as lie at the heart and marrow of things." Elsewhere he states, "Experiments of this kind have one admirable property and condition: they never miss or fail. Since they are applied, not for the purpose of producing any particular effect, but only of discovering the natural cause of some effect, they answer the end equally well which ever way they turn out; for they settle the question."

Bacon's concepts of scientific research seem naive to us because experience has shown that few important discoveries have been made by applying the purely inductive method as mechanically as he advocated. William Harvey, who was Bacon's physician, was aware of the scientific naiveté of his illustrious patient and dismissed his importance as a scientist with the remark, "He writes Philosophy [meaning science] like a Lord Chancellor." Harvey's remark expresses the scorn and irritation that most experimenters feel toward those who express opinions on scientific subjects from second-hand knowledge, but it betrays also a failure to recognize the considerable influence that Bacon's writings would have on social evolution. Bacon recognized his limitations as an experimenter and considered his role that of a gadfly; in his own words, he was the one who "rang the bells which called the wits together."

Although the experimental method has now been used continuously for more than 300 years the findings of theoretical science did not begin to affect human life in a significant manner until the second half of the nineteenth century. Before that time most successful inventors were empiricists who depended on their wits and manual skills rather than on theoretical concepts. The famed twentieth century Harvard chemist and sociologist L. J. Henderson was fond of remarking that before 1850 the steam engine had done more for science than science had done for the steam engine. This witticism expresses the historical truth that for most of human history theoretical knowledge has contributed little to technology and has in fact been largely dependent on empirical inventors for information and instrumentation. Countless technological achievements have emerged independently of theoretical science yet have profoundly influenced human life and science itself. They range from the Sumerian clay tablets to Gutenberg's printing machine, from the water wheel to Watt's steam engine, from the distillation of alcohol to the use of legume plants as fertilizers.

Until the past few decades the differentiation was so profound between theoretical knowledge and the practices of agriculture, medicine, and technology that true scientists were inclined to look down upon inventors. Even Michael Faraday, whose experiments generated so much modern industry, preferred to be called a natural philosopher rather than a man of science, probably because he wanted to be identified with noble and eternal truths rather than with commercial applications of transient value.

An amusing example of the snobbery of scientists toward inventors is found in the remarks made by the English theoretical physicist James Clerk Maxwell while lecturing on the invention of the telephone by Alexander Graham Bell in 1878:

When, about two years ago, news came from the other side of the Atlantic that a method had been invented of transmitting, by means of electricity, the articulate sounds of the human voice, so as to be heard hundreds of miles away from the speaker, those of us who had reason to believe that the report had some foundation in fact began to exercise our imaginations in picturing some triumph of constructive skill—something as far surpassing Sir William Thomson's Siphon Recorder in delicacy and intricacy as that is beyond a common bellpull. When at last this little instrument appeared, consisting, as it does, of parts, every one of which is familiar to us, and capable of being put together by an amateur, the disappointment arising from its humble appearance was only partially relieved on finding that it was really able to talk.

Later on in the lecture Maxwell stated, "Now, Professor Graham Bell, the inventor of the telephone, is not an electrician who has found out how to make a tin plate speak, but a speaker, who, to gain his private ends, has become an electrician."

The inventors, on the other hand, were themselves often derisive toward the theoretical mathematicians and impractical laboratory men. Had they read Francis Bacon, they would have welcomed the following statement by him:

> Although the roads to human power and to human knowledge lie close together, and are nearly the same, nevertheless . . . it is safer to begin and raise the sciences from those foundations which have relation to practice, and to let the active part itself be as the seal which prints and determines the contemplative counterpart.
>
> Celsus . . . tells us that the experimental part of medicine was first discovered and that afterwards men philosophized about it, and hunted for and assigned causes; and not by an inverse process that philosophy and the knowledge of causes led to the discovery and development of the experimental part. And therefore it was not strange that among the Egyptians, who rewarded inventors with divine honors and sacred rites, there were more images of brute than of men; inasmuch as brutes by their natural instinct have produced many discoveries, whereas men by discussion and the conclusions of reason have given birth to few or none.

The French Encyclopedists also sided with the inventors against theoreticians. When, in 1751, they undertook the publication of their *Universal Dictionary of Arts, Sciences, Trades, and Manufactures*, they devoted much space to technical processes as carried out in workshops. "Should not," they asked, "the inventors of the spring, the chain, and the repeating parts of a watch be equally esteemed with those who have successfully studied to perfect algebra?"

The distinction between the theoretical aspects and practical applications of science has become increasingly blurred during the past 150 years. Faraday, natural philosopher though he claimed to be, in fact moved in and out of practical fields throughout his life. Pasteur expressed an attitude now widely held when he wrote, "There are not two sciences. There is only science and the applications of science and these two activities are linked as the fruit is to the tree." (Quoted in Dubos, 1950)

Many scientists and historians of science, however, still hold to the distinction between pure science and applied science. The English mathematician G. H. Hardy endorsed this distinction when he offered his famous toast: "Here's to pure mthematics. May it never have any use." In a more serious mood, it is often claimed that the basic difference between science and technology is one of intention: science tries to explain, technology to construct. The following quotation presents a more elaborate statement of this view: "By science, I mean the search for knowledge and understanding—both the understanding of something that is complex in terms of its simpler components and the understanding of a given phenomenon in terms of the relations between it and our other knowledge about the world. By technology I mean the application of whatever is presumed to be already known to the accomplishment of immediate goals." (Hirsch, 1967)

In practice most contemporary scientists engage at times in scientific theory and at times in the development of scientific practice. Technologists find it essential to keep in touch with theory, if only to derive ideas that give them an advantage over their competitors. Practicing physicians know that diagnostic, preventive, and therapeutic procedures are continuously being transformed by medical science. Theoreticians, on the other hand, can hardly escape contact with the practical applications of their science because so many of them act as advisors to government and consultants to industrial or commercial firms. Practically all scientists today follow the path opened by Faraday, Kelvin, Liebig, Pasteur, and many other luminaries of the nineteenth century who acted sometimes as natural philosophers, sometimes as inventors, passing from theoretical studies to practical applications as circumstances dictated.

Science, either in its theoretical or applied forms, is practiced by many different types of human beings so different that it is impossible to find a characteristic common to all of them.

For example, the social anthropologist who goes to live among primitive people to study their sexual mores engages in a kind of scientific work best carried out by someone whose personality is different from that of the organic chemist who works on the molecular structure of the odoriferous substances responsible for sexual attraction or the statistician who studies the mathematical relations between lunar cycles and sexual activity.

Nor does the scientific enterprise derive unity from the kind of intellectual endowment required of its practitioners. Curiosity is of course a *sine qua non* attribute of scientists. According to H. L. Mencken the prototype of the scientist "is not the liberator releasing slaves, the good samaritan lifting up the fallen, but a dog sniffing tremendously at an infinite series of rat holes." Curiosity suggests questions that each individual tries to answer according to his own temperament, but the same could be said for all other types of occupations.

The practice of the experimental method is compatible with as many attitudes as there are experimenters. The accounts that famous scientists have given of the manner in which they made their discoveries show that the solution of a problem may come in many ways: it can result from conscious, laborious and intense systematic analysis; it may appear as a sudden vision; it can occur during a pleasant country walk or while asleep; or it can come through images and analogies while daydreaming by a fireplace or in a crowded city bus.

The thirst for knowledge is of course a very desirable attribute of the scientific temperament, but it often takes the form of intellectual lust rather than of a search for truth. In his novel *Arrowsmith*, Sinclair Lewis wrote: "Most people who call themselves truth seeks do not so much desire to find the truth as to cure their mental itch." Indeed, many investigators have acknowledged that much of the pleasure derived from a discovery or other scientific achievement is similar to that experienced in overcoming any kind of difficult task. "The burning desire for knowledge is what motivates and supports the efforts of the investigator," wrote Claude Bernard; but immediately he added, "The fact that knowledge endlessly recedes as the investigator is about to grasp it, is what constitutes at the same time his torment and his happiness." Max Planck stated in a similar vein, "It is not the possession of truth, but the success that attends the seeking after it, that enriches the seeker and brings happiness to him."

More often than not the pursuit of science is thus akin to an intellectual sport and yields rewards which are independent of the specific nature of its goals. In this respect again, scientific research does not differ greatly from other human enterprises. Cervantes asserted that the road is always better than the inn, and Robert Louis Stevenson claimed that to travel hopefully is better than to arrive safely.

Nor is there any reason to believe that the practice of science requires an intellectual discipline or moral attributes different from those found desirable in other human activities. A few years ago a professor of education made a survey among sixteen leaders in different fields of the natural sciences to learn from them what characterized the scientific attitude. The replies gave a dominant place to open mindedness, inquisitiveness, intellectual curiosity, intellectual honesty, objectivity, industry, perseverence, critical independent reflection, and a willingness to label conclusions tentative until supported by reproducible and convincing data. This is not a very original list nor one that is of any help in defining the scientific attitude. An honest person in any field would insist on being open-minded, objective, industrious and on considering all the evidence before reaching a conclusion. Scientists have no monopoly on the rational processes of thought. In fact, they do not use the scientific method outside a very narrow aspect of their professional specialization, not even in their social behavior during their professional activities, let alone in their personal lives.

The mental attributes of the perfect man of science were defined as follows 300 years ago by Thomas Sprat, Bishop of Rochester, in his *History of the Royal Society of London* (1667):

If I could fetch Materials whence I pleased to fashion the Idea of a perfect Philosopher, he should not be all of one Clime, but should have the different Excellences of several Countries. First, he should have the Industry, Activity, and inquisitive Humour of the Dutch, French, Scotch and English, in laying the Ground Work, the Heap of Experiments; and then he should have added the cold, circumspect, and wary disposition of the Italians and Spaniards, in meditating upon them, before he brings them fully into Speculation. All this is scarce ever to be found in one single Man; seldom in the same Countrymen.

Thomas Sprat's statement may not be of much help in defining the qualities required for the practice of science. But it has the merit of warning us that their prevalence in a given country is not as permanent as is commonly thought. National characteristics can change drastically in 300 years or even in a much shorter time.

The word "scientist" is of recent origin. As far as is known, it does not appear in print either in English or any other language until 1841. Before that time the student of natural phenomena was called a "Man of Science" or a "Natural Philosopher." As late as 1895, according to the British science analyst Ritchie Calder (now Lord Ritchie-Calder, 1964), the London *Daily News* was still objecting to the use of the word "scientist" as "this American innovation." To his dying day H. G. Wells insisted that the proper term was "Man of Science."

In the past the man of science tended to be intellectually and socially somewhat broader than the modern scientist. James Watt, the inventor of the steam engine, argued music with Herschel the Astronomer Royal, who had been a German band master. William Small, who was Professor of Natural Philosophy at Williamsburg, taught Thomas Jefferson the checks and balances of Newtonian physics, which were transferred to the American Constitution in the form of counterweights of the Executive, Legislative and Judiciary branches. Helmholtz applied his knowledge of optics to the appreciation of painting and his knowledge of acoustics to the understanding of harmony in music. Louis Pasteur argued with the philosopher Ernest Renan about the relative merits of experimental science and the historical sciences. (Dubos, 1950)

Until this century most men of science could converse with any other educated man; they were specialized in their skills but not in their interests. Although their trust in the power of experimental science had caused them to retreat from scholasticism, they had retained a scholarliness that enabled them to maintain contact with the humanities.

During the second part of the nineteenth century men of science began to turn into scientists. The "ist" in the new word meant that instead of being natural philosophers on speaking terms with their colleagues in the humanities, scientists were becoming specialized in their attitudes as well as in their fields of knowledge. They were concerned primarily and often almost exclusively with the problems and techniques peculiar to their scientific specialty.

The reason most commonly given for the shift from natural philosopher or man of science to scientist is that the increase in scientific knowledge made it necessary for one to focus on a special field and consequently neglect other areas of interest. I doubt the validity of this explanation. Among scientists who died during the past few years, Erwin Schrodinger, Robert Oppenheimer, and Werner Heisenberg, were illustrious physicists, yet each had much of importance to say concerning philosophical and social problems.

Another reason given for the disappearance of the natural philosopher is that the educated men of the nineteenth century were members of a small elite, whereas almost anyone now can become a scientist. This also is questionable. Many of the most famous and best educated men of science in nineteenth century Europe came from the laboring classes or the small bourgeoisie—as was the case with Faraday and Pasteur. The obituaries of philosophically minded members of the National Academy of Sciences in the United States reveal that a very large percentage were of humble origin, born and brought up in small towns or on farms. Now, as in the past, social origin has little if anything to do with the breadth of one's interests.

In my opinion the intellectual narrowness of many specialists (and this remark does not apply only to the specialists in the natural sciences) comes from the widespread assumption that the discovery of new facts is the most important aspect of knowledge. This assumption is illustrated by what has happened to the academic programs for Ph.D. training. If one really believes that the *advancement* of knowledge is more important than the *possession* of knowledge, then it may be justifiable to limit one's attention to the kinds of skills and facts required for technical progress in one's particular field, whether Mesopotamian ceramics or wave mechanics. Advanced academic programs pay lip service to the need for knowledge in areas different from the field of specialization of the candidate. In most cases, however, this requirement is just a façade; the so-called doctorate in philosophy is now a misnomer, a certificate of expertise in a narrow specialty rather than of philosophical understanding or even awareness of the interrelationships among the various fields of knowledge.

Ortega y Gasset was obviously exaggerating when he asserted in *The Revolt of the Masses* (1932) that "science automatically converts the scientist into a modern barbarian." He was nevertheless pointing to a common weakness in the scientific community when he wrote, "The specialist . . . is not learned, for he is formally ignorant of all that does not enter into his specialty; but neither is he ignorant, because he is a 'scientist' and 'knows' very well his tiny portion of the universe. We shall have to say that he is a learned *ignoramus*."

It is no consolation that Ortega should not have limited his criticism to training in the scientific disciplines, because the "learned ignoramus" is just as familiar in the humanities as in the sciences. What must be emphasized is that accumulated "public knowledge," both in the humanities and in the sciences, is as important as individual "research" achievements. In part, our elaborate institutional apparatus for the discovery of new facts serves only "to add a few more pages to the books on the shelves." (Ziman, 1968) The health of the learned community depends as much on the savant as on the specialized investigator.

The expressions scientist, science, and scientific method thus have meanings which are poorly defined. The classifier of beetles is called a scientist; yet he differs from the theoretical physicist as much or perhaps more in his activities than he does from the student of Indo-European languages, who is called a humanist. Man engages in a kind of activity that can be called science whenever he tries to describe objects and phenomena, to analyze each in its component parts, to discover interrelationships, and to convert knowledge into some form that is of use to mankind. This definition applies equally well to studies concerning the fundamental units of matter or energy, the myriads of living creatures that have arisen in the course of evolution, the components of man's body or brain, and also the behavioral and social problems about which reliable and reproducible information can be obtained.

In theory it is difficult to define where science ends and where the humanities begin. The lay public, however, has no difficulty in differentiating between the two. Non-science books, concert halls, and art museums have more popular appeal to the lay public than do books or exhibits about science. Scientists themselves react like the lay public when they function outside their areas of professional specialization. The student of plasma physics or of plasma proteins is not likely to select books on marsupial anatomy for his bedside reading, nor is the organic chemist inclined to become familiar with problems of population genetic polymorphism. Most scientists, it is true, are interested at present in radiation fallout and in the hidden surface of the moon, but so are many members of the Chamber of Commerce. Winston Churchill, Pablo Picasso, and Ernest Hemingway are much more frequently discussed at the luncheon tables of scientific research institutes than are the Nobel Prize winners in physics, chemistry, or biology of the same generation. And if the name Robert Oppenheimer or Linus Pauling is mentioned, the respective achievements of these investigators as physicist or chemist are less likely to be emphasized than their behavior as interesting and vital human beings. While scientists qua scientists are deeply committed to the development of their own specialized fields, they generally turn to nonscientific topics when they move outside their professional spheres.

Whatever historians and philosophers of science may say concerning the fundamental similarities between science and the humanities as intellectual and creative pursuits, the high school or college student soon discovers from his personal experience that the two kinds of learning and activities are different as far as he is concerned. He will probably like one and dislike the other; and science too often loses in the comparison.

The intellectual effort required in the study of most scientific subjects accounts in part for their neglect by the lay public and by most young people, but there is more to it than that. Literature and the arts, when they are meaningful to the common man, deal with the world in which people are born, live, and die; they both portray situations in which the human being is completely engaged and responds in a manner that constitutes the real experience of life with its confusing but vital richness. They entice the senses as much as the rational mind; in place of proof they give possession, and in place of averages they give wholeness. They deal with situations that concern the living experience in which man can recognize himself. In contrast, all sciences, whether theoretical or applied, depend upon analysis; they are concerned almost exclusively with those aspects of experience which are the least personal.

In his William James Lectures at Harvard University, Etienne Gilson stated that "Every scientist naturally has the temper and the tastes of a specialist . . . the natural tendency of science is not towards unity, but towards an ever more complete disintegration." Gilson's statement is only a half truth. The more imminent danger is that the specialist commonly loses contact with the aspect of reality which was his primary concern, whether it was matter, life, or man.

In his own experience of the physical world the physicist does not use his specialized knowledge for a richer or more subtle contact with reality; nor is the biologist rendered capable of perceiving the living experience more acutely because he is familiar with intermediate metabolism or X-ray diffraction patterns of contractile fibers. The theoretical physicist apparently finds it difficult to convert the mathematical formulae on which he depends for his work into experiences or thoughts meaningful to his own senses and reason. The general biologist finds no trace of the creativeness of life in the macromolecules he isolates from the cell. The student of consciousness cannot relate the operations of the isolated sense organs or nerve impulses to the emotions elicited by a fragrant rose or a romantic sunset.

There has been much talk during recent years of the lack of communication between the humanistic and scientific aspects of knowledge. In reality, however, this disjunction is not so critical as is often suggested. As Oppenheimer stated, "Occasionally between the sciences, and more rarely between a science and other parts of our experience and knowledge, there is a correspondence, an analogy, a partial mapping of two sets of ideas and words. We learn then to translate from one language into another. Ours is thus a united world, united by countless bonds." (Quoted in Rabi, et al., 1969)

Each of us can and does learn many facts and concepts pertaining to areas of knowledge totally different from the one in which he is a specialist. The breakdown in communication is complete only when the concepts cannot be related to direct human experience. The physicist, the biologist, the humanist, and the layman can all find a common ground for discourse if they talk about matter, life, or man as perceived by the senses or as apprehended in the form of images, analogies and responses. But discussions of matter in terms of mathematical symbolism or a view of life and man in terms of disintegrated components cannot be related to any form of direct human experience. Specialists must return to the original human basis of their work if they want to converse with their fellow men.

3
Visions and Disenchantment

The Scientific Dreams of Mankind

While Benjamin Franklin was Ambassador to the court of France, he had occasion to witness in Paris some of the first balloon ascents. To skeptics who wondered what use a balloon might ever have, Franklin is reputed to have answered, "What is the use of a newborn baby?"

A similar story is told of Michael Faraday. Shortly after he had discovered electromagnetic induction, but before it had been converted into a practical technology, Faraday received a visit from an important political personage in his laboratory at The Royal Institution in London. He demonstrated the new phenomenon to the visitor who was not impressed by Faraday's simple apparatus and inquired, "What is the good of this discovery?" Faraday replied, "Some day, sir, you will collect taxes from it."

Whether or not these two stories are authentic is of limited historical interest, but it is significant that they were told early in the nineteenth century. They illustrate that faith in the power of scientific and technologic research was already prevalent in Western Europe before the Industrial Revolution transformed human life.

Franklin's facetious remark, "What is the use of a newborn baby?" implies that any discovery holds the promise of unforeseeable theoretical and practical developments. Faraday's assertion that his new technique, as yet but a laboratory stunt, would become a source of taxes for the state suggests that most scientific discoveries lead to processes or products that the public is eager to use.

The confidence voiced in Franklin's and Faraday's quips now seems commonplace because it has been substantiated by countless demonstrations of the practical usefulness of scientific research during the past hundred years. But there was no factual basis for this confidence until the middle of the nineteenth century. Before that time science was the avocation of natural philosophers and had not yet greatly influenced the practical affairs of man. Franklin and Faraday shared with their contemporaries the unproven belief inherited from Bacon that knowledge and understanding would eventually enable man to transform his life through his control of natural forces.

Belief in the possibility of understanding and controlling nature constitutes the very basis of technological civilization. This belief emerged among some ancient peoples and persisted even during the unproductive periods of history; but it was a faith that remained unsubstantiated until recent times.

Among the ancient peoples who believed they could understand nature, the Hellenic philosophers collectively shared the faith that it is possible to arrive at a rational explanation of the cosmos; individually each had faith that his own understanding of the cosmos was the correct one. Heraclitus believed that everything is forever in flux, that reality is movement and change; in contrast Democritus believed that everything is made up of hard particles, which he called atoms; Phythagoras believed that the observed universe reflects an inherent orderly arrangement of a numerical character; Aristotle believed that all parts of the cosmos are arranged according to a predestined system determined by final causes. But the Hellenic scholars did not have any factual knowledge to verify these beliefs; their conceptual views of rationality and the world were merely founded on faith.

In every age mankind has experienced a number of collective dreams which, albeit differing in form from one culture to another, reveal the same fundamental preoccupations everywhere and always—as if they were Jungian archetypes. These collective dreams have provided many of the most entrancing myths, legends, and tales of humanity; some of them are now being converted into reality by modern science.

For example, the Icarus myth symbolizes the ancient dream of flying; Leonardo da Vinci's design of flying machines is but one of the many attempts to apply existing technological knowledge to the realization of this dream. Since antiquity, poets, storytellers, and science fiction writers have given imaginative descriptions of ways to reach the planets and the stars and have speculated as to the kind of life to be found there. The existence of space rockets was imagined long before they could be conceived in operational terms, and Jules Verne was far-sighted enough to locate the launching site of his moon vehicle in Florida, quite near Cape Kennedy.

The legends associated with ancient temple knowledge, the mysterious learning and processes of the alchemists, and the obscure Rosicrucian beliefs point to the fact that some of the most modern pursuits of science have been in human consciousness for more than 2,000 years. Such problems as the transmutation of elements, the power to see what is happening far away, the discovery of hidden substances, the transmission and the control of thought, the fountain of youth, and indeed the creation of life from inanimate matter are a few of these scientific pursuits.

Various forms of collective and individual faith have continued to dominate science until modern times, and there is no indication that human attitudes are changing in this respect. Many inventors still behave like the alchemists who labored over smoky retorts in ill-smelling and cluttered cells, endlessly trying to manufacture gold from other elements and to discover the elixir of life.

What an extraordinary faith it demanded of the Franciscan monk, Roger Bacon, to predict in the thirteenth century technological developments that were realized long after his death—the circumnavigation of the world, the propulsion of boats by mechanical means, the utilization of the explosive property of powder, the improvement of sight by the proper adjustment of lenses, and even the technology to accomplish flight through the air. Four centuries later, Francis Bacon described in his last book, *The New Atlantis* (1627), an utopian community of scholars that he called Salomon's House. Through scientific research the scholars had discovered and invented many things we take for granted today: flying machines, submarines, instruments "for hearing at a distance," synthetic drugs, and perfumes. The activities of Bacon's imagined scientists were focused on the practical applications of knowledge for the "benefit and use of man, the relief of man's estate."

It was also on purely unsubstantiated faith that Francis Bacon and the philosophers of the Enlightenment believed that the systematic application of science to the affairs of man would bring prosperity, health, and happiness.

Franklin, matter of fact though he was, is again worth quoting regarding his faith in the benefits of science. In a letter to his English friend, the chemist Joseph Priestley, he was so euphoric as to state: "It is impossible to imagine the height to which may be carried, in a thousand years, the power of man over matter. We may perhaps learn to deprive large masses of their gravity, and give them absolute levity, for the sake of easy transport. Agriculture may diminish its labor and double its produce; all diseases may by sure means be prevented or cured, not excepting even that of old age, our lives lengthened at pleasure even beyond the antediluvian standard."

A recent anthology of writings by Erasmus Darwin (King-Hele, 1968), who lived in the eighteenth century and was Charles Darwin's grandfather, provides even more extraordinary examples of the human mind's ability to forecast the shape of things to come—from the role of phosphorus and nitrogen in fertilizers to the theory of Artesian wells and the evolutionary importance of sexual reproduction.

The scientific dreams of mankind do not, of course, constitute science either in its theoretical or practical aspects. Nevertheless, these dreams may have accelerated scientific development by providing a formal structure for the scientific effort.

During the hundred years between Dalton and Heisenberg the atomic theory was regarded by many as the most profound and fundamental truth about Nature; so that any man found proclaiming it more than two thousand years before Dalton seemed to be a scientific genius. Since the development of wave mechanics, the situation is somewhat changed. Classical atomism is dead, and theoretical physicists can discuss quite seriously whether so-called 'fundamental particles' might not be replaced by mathematical singularities in fields of force —a conception having more in common with the continuum theories of the Stoics than with the unvarnished atomism of Democritos. (Toulmin, 1968)

As soon as the state of knowledge indicated that

anticipatory imaginings could be converted into some form of reality, scientists have rushed to the study of certain problems which had long appeared unscientific or even antiscientific. For example, the transmutation of elements, long postulated by alchemists and Rosicrucians, became a flourishing enterprise when theoretical physics provided the proper scientific background. The grafting of organs was imaginatively depicted by Renaissance painters before surgeons, physiologists, and immunologists began to work out techniques for making it a reality.

In subconscious ways the dreams of mankind may still be influencing the orientation of the scientific effort. A remarkable kind of faith has been demanded in our own times for such a large amount of scientific talent and financial resources to be devoted to the creation of giant accelerators and space craft for which there is no urgent practical use in sight. This has occurred at a time when even the most affluent societies are faced with immense unsolved problems among their deprived peoples and when the destruction of the world by nuclear warfare appears to be a real possibility.

It was also faith which motivated the designers of the World Health Organization charter to affirm that health is the birthright of man and to regard health not merely as the absence of disease but as a positive attribute of complete physical and mental well-being. Such a statement demands much confidence in the essential sanity of man's nature, when it is known that countless people all over the world are malnourished and suffer from infection, that there is no hope of being able to provide adequate medical care in most places, and that new diseases appear as the old ones are rooted out. This kind of faith, unreasonable when first expressed, has generated some of the most effective health programs in the underprivileged parts of the world. As George Bernard Shaw once remarked: "The reasonable man adapts himself to the world; the unreasonable one persists in trying to adapt the world to himself. Therefore all progress depends on the unreasonable man."

The scientific dreams of mankind have also taken the form of large philosophical concepts. For instance, Lovejoy's famous book *The Great Chain of Being* (1936) illustrates the universality and antiquity of the belief that all living things are arranged according to an orderly natural system, referred to in late medieval times as the Great Ladder of Being. This belief emerged repeatedly in the philosophical and religious doctrines of both the distant and recent past before becoming integrated in orthodox science through evolutionary and genetic theories. At present, the popular appeal of Teilhard de Chardin's science and mysticism probably has a similar basis in man's emotional need for relating himself to the rest of creation.

Even the most ancient societies have been concerned with the origin of the cosmos, either believing that it came into being through a special act of creation, or that it has always existed in a form similar to the one perceived by the senses. The modern versions of these two ancient and still popular concepts were discussed in the concepts of physics by Fred Hoyle, the Professor of Astronomy at Cambridge University, when he delivered the Bakerian lecture before the Royal Society in June, 1968. According to the Big Bang Theory the universe was created at a given time by expansion of matter from a more tightly compressed state. Fred Hoyle, however, favors steady-state cosmologies, in part for reasons based on theoretical physics, but also for philosophical reasons. Why, he asks, should the primordial atom blow itself up at one given instant in eternity? Is it not more reasonable to envision a continuing state of affairs in which matter is constantly being replaced by the spontaneous creation of new matter? The Big Bang Theory calls for a "singularity" in time which Fred Hoyle considers at variance with scientific thinking. He poses the question: "Do singularities exist in the real world, or are they metaphysical entities?" Clearly these are questions which transcend orthodox experimental science and express philosophical preoccupations as old as mankind.

Very ancient, also, is the faith that some form of underlying unity pervades the Universe. This faith was a vague philosophical concept until a few decades ago, but now it has become a motivating force among theoretical physicists. I would not presume to discuss a problem so obviously beyond my comprehension, but I might be permitted to quote here a few statements made by Werner Heisenberg, one of the adherents of this controversial doctrine, during an interview published in *The New Yorker Magazine.* "I am convinced that the Universe is connected by a truly simple law," Professor Heisenberg is reported to have said. To perceive "the sense of harmony that underlies everything . . . we must see the big connection. To understand nature we must make order out of phenomena."

If one were to judge from the writings of the late Alexandre Koyré, one of the most respected students of the history and philosophy of science, this order originates chiefly from man's own conceptual view of "seeing" the world: "Good physics is made *a priori.* Theory precedes fact. Experience is useless because before any experience we are already in possession of the knowledge we are seeking. Fundamental laws of motion (and of rest), laws that determine the spatio-temporal behaviour of material bodies, are laws of a mathematical nature. Of the same nature as those which govern relations and laws of figures and of numbers. We discover them not in Nature, but in ourselves, in our mind, in our memory, as Plato long ago has taught us. . . ."

Needless to say, I am not quoting either Alexandre Koyré or *The New Yorker* as authorities on problems of modern theoretical physics. What I wish to emphasize is Heisenberg's statement that "We must *make* order out of phenomena" because this expresses so forcefully how science is used to provide a rational structure for collective beliefs that have deep roots in the mental substratum of *Homo sapiens.*

Visions and Disenchantment

The Nightmares of Reason

One of Goya's most famous etchings shows a scholar sleeping at his desk. A weird-looking cat and several bats surround him as in a nightmare. The desk bears the inscription "El sueno de la razon produce monstros," which can be translated, "The sleep (or dream) of reason generates monsters." (Dubos, 1961)

The creatures in the etching probably symbolize the bestial acts of war and civil strife that paradoxically marked the Age of Reason in Europe. They express Goya's anguish at seeing monstrous acts being committed in the name of rationality. Science without conscience is but death of the soul, Montaigne had written in the 16th century. Reason can become destructive when it is not guided by worthy human concerns.

Reason has been immensely productive of scientific knowledge and technological achievements since the Renaissance. But it must have suffered from intoxication and nightmares during the past few decades if one judges from the technological monsters it has engendered.

Scientific knowledge could certainly liberate mankind from grinding poverty and enlarge our view of the universe and of life. In fact, until recent years, all scientists and most laymen had taken for granted the assumption that science was inherently good and that the technologies derived from it constituted the major agents of desirable change and social progress. This view was challenged by only a few humanists and romantics unrealistically advocating a return to the good days of the past. The public mood is now changing, as evidence accumulates that technology is forging new shackles for man from which he seems unable to escape. Nuclear weapons threaten to destroy all living things; industry and transport are polluting the world with waste products, visual insults, and deafening noises; mass media are conditioning and cheapening mental and emotional responses; and microelectronics are invading all aspects of life and making privacy almost impossible. The prodigious achievements of modern technology seem to be paving the way for a new kind of totalitarianism.

The uneasiness of the public about the present state of affairs and the consequent apprehension about the future are reflected in the topics selected by science fiction writers and in the popularity of anti-utopian literature.

Science fiction has been defined as that form of literature which, growing with science, evaluates it and relates it meaningfully to the rest of existence. Furthermore, science fiction constitutes a kind of psychotherapy, because imaginings about the past or the future help release man from the grip of the present.

Fanciful descriptions of travels to the outer world have long been a popular form of escape from unpalatable reality. They began with early Greek writers and have continued with Plutarch, Cyrano de Bergerac, Jonathan Swift, Jules Verne, H. G. Wells, and a legion of contemporary science fiction writers of today. Space vehicles, imaginary or real, have always been loaded with the protestations, aspirations, and illusions of mankind.

The most popular scientific book for the general public published during the eighteenth century was probably Bernard de Fontenelle's *Entretiens sur la pluralité des mondes* (1686). This book was not science fiction as we understand it today but was an intelligent and entertaining discourse about humanistic aspects of science. It was written for a fashionable public which wanted or pretended that it wanted to be enlightened; in any case, it was essentially unconcerned with the social implications of knowledge.

Socially oriented science fiction emerged during the nineteenth century. Like the great international expositions of that time it reflected the euphoria of the general public about the potentialities of science and of technology. A review of American science fiction in the nineteenth century has been published recently under the title *Future Perfect* (Franklin, 1967); the stories discussed have an optimistic tone because they all take for granted that science and technology will solve mankind's problems, provide health, abundance, and excitement for all, and generally raise the level of happiness. Its title, *Future Perfect*, symbolizes the mood of scientific utopias.

In contrast, two other studies of science fiction, also published recently, are entitled *New Maps of Hell* (Amis, 1960) and *The Future as Nightmare*

(Hillegas, 1967). They deal chiefly with mid-twentieth century science fiction and analyze in particular the flood of anti-utopian literature which has appeared since the Second World War. Whereas the mood of *Future Perfect* is naturally optimistic, the dominant themes in *The Future as Nightmare* and *The Maps of Hell* are the catastrophic events that will inevitably result from nuclear warfare and the horrid forms of life that will prevail in the future. The same anti-utopian mood is conveyed in another collection of stories under the apt title *The Case Against Tomorrow* (Pohl, 1957).

More interesting than the accounts of destruction by nuclear bombs are the pictures of technology over-reaching itself and destroying society. Jules Verne foresaw this danger more than one hundred years ago and portrayed it in "Five Weeks in a Balloon" (1862): . . . "If men go on inventing machinery they'll end by being swallowed up by their own inventions. I've often thought that the last day will be brought about by some colossal boiler heated to three thousand atmospheres blowing up the world."

"And I bet the Yankees will have a hand in it."

In the story "The Midas Plague," from the collection *The Case Against Tomorrow*," Frederik Pohl deals with the social consequences of life in a world where the "tireless labor of humanity and robots drove jungle and swamp and ice off the Earth, and put up office buildings and manufacturing centers in their place." This is a world in which the hero had nothing to do but "to eat and drink and wear and wear out his share of the ceaseless tide of wealth." A more sophisticated view of the dangers posed by the technological take-over is found in Ray Bradbury's *Fahrenheit 451*. In this novel man's longing to abandon reality is fulfilled through the widespread use of mechanical wonders, which promote quiescence at the cost of neurosis.

So much anti-utopian literature has been published during recent decades that reaction against it is beginning to take place. The new forms of utopian scientific literature stimulated by this reaction, however, have a mechanistic optimism which is as chilling as the pessimistic anti-utopian mood they are trying to counteract.

In his utopian novel *Andromeda*, the Russian writer Ivan Yefremov describes a Great World State organized according to orthodox Soviet values (Hillegas, 1967). In the present capitalistic world, according to Yefremov, science is used for torture, punishment, and thought reading, as an instrument of intimidation "to turn the masses into obedient semi-idiots, ever ready to fulfill the most monstrous orders." In contrast, the citizens of the Great World State depicted in *Andromeda* have learned to make knowledge contribute to happiness by putting science and technology at the service of mankind, along socialist lines. Individual egos are no longer a problem in *Andromeda* because the utopians are educated and conditioned from childhood to master their selfhood and develop a high degree of social conscience.

In *Walden Two* (1948), B. F. Skinner describes

paradise on earth as conceived by the behavioral psychologist. Men are scientifically conditioned to be good; competitiveness and aggressiveness are eliminated through the proper psychological upbringing and manipulation of behavior. The slogan of *Walden Two* is "Our education is our government." Using the techniques of behavioral engineering, especially positive reinforcement, the social elite of *Walden II* rules for the greatest happiness of all by making people *want* to do what they are *expected* to do.

Aldous Huxley's *Island* (1962) also portrays a psychological utopia, but it is achieved through means different from those used in the paradise of *Walden II*. In *Island,* man's libido is freed and aggressiveness is ended because men have been emancipated from economic and sexual repression. Aldous Huxley makes it even easier for his Utopians to escape from the constraints of reality by providing them with the yoga of love, the consciousness expanding moksha-medicine, and hypnotism for destiny control.

The reaction against the anti-utopian mood of *The Case Against Tomorrow, The Maps of Hell,* and *The Future as Nightmare* extends beyond the kinds of psychological paradise imagined by B. F. Skinner and Aldous Huxley. This reaction is now taking the form of technological utopias tediously stimulated by the approach of the year 2000. Although the modern technological utopias are described with a complex apparatus of scientific scholarship, their authors can be classified among writers of science fiction because they organize their prophecies in the form of scenarios which are intended to have the vividness of real-life situations.

For example, Herman Kahn and Anthony Wiener of the Hudson Institute have recently published a book entitled *The Year 2000* (1968). The institute is a non-profit research center at Croton-on-Hudson, New York, that makes predictive studies for the government, industrial firms, and other private organizations. The book describes the scientific, technological, and medical advances that can be expected to shape life by the year 2000. A more recent book in the same spirit is entitled *The Year 2018* (1968). Its publication was sponsored by the Foreign Policy Association to celebrate its fiftieth birthday in 1968, and was presumably published under the illusion that the book will help the members of the Association to visualize how the world will look in the future.

As could be expected the two books predict a number of spectacular breakthroughs in the production of nuclear energy, the development of new electronic gadgets, and the synthesis of strange chemical products. Of the one hundred specific predictions in *The Year 2000,* I shall list here only a few among those that are claimed to have a direct bearing on human life.

—Permanent manned satellites and lunar installations;

—Interplanetary travel;

—Permanent inhabited undersea installations, and perhaps even colonies;

—Artificial moons and other methods of lighting large areas at night;

—Robots and machines "slaved" to human beings;

—New, more varied, and more reliable drugs for control of fatigue, mood, personality, perceptions, and fantasies;

—Increase in life expectancy, postponement of aging, and limited rejuvenation;

—Extensive use of mechanical aids or substitutes for human organs, senses, and limbs;

—Permanent cosmetological changes (features, perhaps complexion, skin color, even physique);

—Human hibernation for relatively extensive periods (months to years);

—Programmed dreams;

—Non-harmful methods of overindulging (whatever that may mean!!);

—More reliable "educational" and propaganda techniques for affecting human behavior (public and private);

—New, pervasive techniques for surveillance, monitoring, and control of individuals and organizations.

In addition to the one hundred "very probable" breakthroughs in the next thirty-three years, the Hudson Institute scientists have also listed a set of developments described as "far out" but still within the realm of possibility by the year 2000.

—Establishment of bases or colonies on the moon or other planets;

—Increase of human mental capacity by connecting the brain directly to a computer;

—Lifetime immunization against practically all diseases;

—Life expectancy well beyond one hundred years.

The authors of *The Year 2000* and *The Year 2018* are careful to emphasize that these lists do not constitute actual predictions of the future; rather, they are statements of what they regard as "falling within the range of scientific probability." Other scientists would doubtless have different views as to which scientific possibilities will be realized by the year 2000 or 2018. But all would agree that, barring natural catastrophes or social upheavals, scientific technology will soon provide new and powerful techniques for manipulating the external world and man's nature.

In view of the resources that will become available if only a small percentage of the predictions now made by scientists become a reality, one might assume that life during the twenty-first century will be safe, comfortable, and exhilarating, at least in the prosperous countries of the industrialized world.

Yet, despite all the modern scientific miracles and the promises of many more to come, the "new pessimism" is prevalent in the most successful technological societies. One possible reason for this paradox is that the technological utopias now being forecast do not correspond to what people really want. Books such as *The Year 2000* and *The Year 2018* constitute travesties of science because they pretend to deal with practical issues yet fail to face up to the realities and needs of the world. The breakthroughs that they predict are trivial because they do not even suggest approaches to the problems that must be solved if mankind is to survive.

We can certainly develop powerful techniques to control and transform our environment, but our man-made and natural landscapes are inferior to what they were two hundred years ago and are rapidly degenerating still further. Technological proficiency provides us with things and services in nauseating profusion, but the harassed faces and the harsh voices of men, women, and children, especially in affluent social groups, reveal that this kind of prosperity does not generate either happiness or peace of mind. Our sumptuous universities and research institutes accelerate the growth and dissemination of knowledge but do not know how to apply professional expertise to fundamental human needs and aspirations.

Man has always lived in a precarious state, worried about his place in the order of things. In the past he was threatened chiefly by natural forces that he could not control, and he experienced fear because of ignorance of the cosmos and of his own nature. Now threats and fears derive in large measure from science and its technologies, paradoxically the most characteristic products of human reason.

The recital of the dangers that technology is creating for mankind and the emotional and philosophical upsets generated by modern knowledge is now heard so frequently as to be one of the most boring aspects of Western literature. It is a mild relief to find that scientists and technologists themselves are among the most articulate spokesmen of this theme. For the sake of illustration I shall quote here recent statements by three illustrious and influential representatives of the chemical, physical, and technological sciences.

At the time of his inauguration as President of the Massachusetts Institute of Technology, in 1964, Julius Stratton addressed the following warning to the faculty and students of the Institute:

> Diseases of the [social] system are emerging in increasing number; and we must be courageous in recognizing that they are themselves the by-products of our highly technological environment.
>
> Consider the transformation of our cities—the physical and social degradation of large areas—the loss of serenity and beauty. We have never before produced so many cars or such fast airplanes; yet transportation in the United States is rapidly approaching a point of crisis. The shift to automation in industry is accelerating and will have profound effects upon the character of our labor force, upon its training, and upon its security. We are polluting our air and water. The pesticides which we are employing on a mounting scale are a boon to agri-

culture and a threat to the remainder of our natural resources.

In 1967 Elmer Engstrom, the president of the Radio Corporation of America, published in *American Scientist* his speech of acceptance of the Proctor Prize awarded by the Scientific Research Society of America. The following quotation expresses the mood of his message: "The introduction of new technology without regard to all the possible effects can amount to setting a time bomb that will explode in the face of society anywhere from a month to a generation in the future."

In the same issue of *American Scientist*, the astronomer Walter Orr Roberts, then president-elect of the American Association for the Advancement of Science, wrote in a similar vein:

Nearly every advance of science has two faces. One smiles on us and lifts the aspirations of man; the other scowls sternly on all future hopes. For the miracle of the modern automobile there is the rising scourge of carborn air pollution that threatens to choke our Bosnywashes (the giant Boston-N.Y.-Washington megalopolis). The advance of urbanism, made possible in part by the miracles of air conditioning and food transportation, brings us befouled rivers, vanishing privacy, and lives full of strain and tension. For all the miracles that atomic energy has wrought in medicine, industry and power generation, there hangs over us the spectre of nuclear war . . . . We will likely know when the first intercontinental missile of World War III comes, should that happen, in a routine computerized check, on a millisecond time scale, of the inventory of space debris; and the decision to retaliate, to enter total war, will probably be made on computer-based advice.

The July, 1968 issue of *BioScience*, the official organ of the American Institute of Biological Sciences, had three consecutive articles, all of them alarmist, under the titles "Can the World be Saved?", "Can We Prepare for Famine?", and "What Role for the Humanist in These Troubled Times?" (Cole, Archer, Predmore, 1968)

The expressions of concern by famous scientists and technologists about the dangers of uncontrolled scientific technology are now commonplace. Those mentioned above are worth recording only because they come from such eminent and influential leaders of the modern scientific establishment. However, they appear less significant, and certainly less original, when read in the light of history.

Protests against the mechanization of life were not infrequent at the beginning of the nineteenth century. Some writers questioned the wisdom of overemphasizing those aspects of science that could be used to exploit nature. They pointed out that science, unless used wisely, could cause changes that would lead to the dehumanization of life. The bankruptcy of science is an old theme for which there has always been a concerned public, but this public has never been able to change the course of events.

The name "Luddites" evokes the English workers who, around 1810, violently resisted the introduction of machinery that threatened their livelihood. In his novel *Erewhon,* first published in 1872, Samuel Butler described an imaginary society shut off from the rest of the world, one in which the inhabitants had destroyed all machinery in the belief that it had caused their ancestors to prefer material things and weakened them by making them dependent on machines. Samuel Butler was anticipating much of the modern anti-utopian literature when he stated through the voice of one of the *Erewhon* leaders: "From a low materialistic point of view, it would seem that those thrive best who use machinery wherever its use is possible with profit; but this is the art of the machines—they serve that they may rule."

Leo Tolstoy went even further than Samuel Butler; he did not limit his protest to the use of machines, but extended it to most of scientific knowledge. He regarded science as a shallow activity making absurd claims to be the fount of knowledge while contributing, in reality, little if anything to the understanding of man and his problems. In a blistering article entitled "The Superstitions of Science," published in 1898, he formulated criticisms that, despite their exaggerations, deserve reading today because they prophetically foretell some of the debates that are now gaining momentum concerning the neglect of human problems by contemporary scientists:

> Practical victories over nature up to the present, and for a long enough time, only lead to factories which ruin the people, to weapons for destroying human life, to the increase of luxury and license. . . .
>
> It is asserted that the reduction of questions of a

higher order to questions of a lower order explains the questions of higher order. But this explanation is never reached, and all that happens is that, descending in its investigations ever further and further from the most real questions to less real questions, science at last reaches a region wholly foreign to man, . . . leaving all the questions which are really important for man totally unsolved.

In words that call to mind those of Ortega y Gasset in *The Revolt of the Masses* (1932), Tolstoy even attacked the intellectual significance of the problems with which scientists busy themselves: "Men of science study, not everything, as they imagine and affirm, but what is most profitable and easy to study . . . this quality belongs not to science, but to people who are inclined to occupy themselves with trifles, and to attribute to these trifles a high importance."

Tolstoy was of course unfair to scientists, in large part, because he did not realize that their dominant intellectual motivation is to understand inanimate nature or at least to arrange their knowledge of it into some kind of system. He seems to have been unaware that in telling scientists that they "must return to the only wise and fruitful understanding of science, according to which its object is the study of how people *ought to* live" (italics mine) he was ignoring their credo. Since the word "ought" involves value judgments it is outside the scientific domain. To understand the ambiguous and often indifferent attitude of the general public toward science, it is important to realize that Tolstoy was acting as the nineteenth century voice of a long tradition when he wrote, "The chemical constitution of the Milky Way, of the new element helium . . . of X-rays and the like . . . none of this is necessary to me; I need to know how to live."

As was mentioned earlier the lectures which gave rise to this book were presented at Columbia University before the Institute for the Study of Science in Human Affairs. The phrase "science in human affairs" is open to at least two very different interpretations. One has to do with the effects that science and its applications exert on human life, making it safer, longer, and more comfortable—although not necessarily happier. There are probably many persons who wonder whether we have not almost reached the point of diminishing returns regarding scientific applications to human life and who fear that further applications may eventually decrease the quality of life. But everyone agrees that science now constitutes one of the most powerful forces in the management of nature and of human life, both for good or for evil.

The phrase "science in human affairs" can refer also to the possibility of using the scientific method for studying what is peculiar to man, his humanness. Most humanists deny this possibility and I believe that most scientists, deep in their hearts, share their skepticism. This is probably the reason why the scientific establishment tends to neglect or at least diminish the importance of observations and experimental studies directly related to living man. Even when physicians leave the bedside and become scientists they are usually more at ease with molecules and machines than with human beings.

In 1965, on the occasion of its hundredth anniversary, The National Academy of Sciences held a series of meeting in Washington to review some of the most important aspects of scientific endeavor. Prestigious American scientists delivered numerous addresses, all learned and brilliant, on various aspects of scientific knowledge, ranging from the structure of elementary particles and the movement of celestial bodies to the submicroscopic structure of animal tissues and the mechanism of hunger in rats. There was hardly any mention of living man in this sophisticated survey of the scientific endeavor, probably because it was judged that the science of man has not yet yielded information of sufficient intellectual quality to warrant its presentation in the august halls of The National Academy of Sciences.

In addition to theoretical and experimental difficulties historical reasons may be responsible for the failure of experimental scientists to concern themselves with man. When modern science emerged in the seventeenth century its founders deliberately excluded the idea of purpose from their concern. More specifically, they took the position that scientists should deal only with causes because inquiry into purposes does not help in understanding nature.

Whatever its philosophical validity, this pragmatic attitude has obviously been immensely fruitful in the study of inanimate matter. But it may be less well suited to the study of living forms and is certainly insufficient when human problems are under consideration. Man is a goal-seeking creature, and purpose is an essential factor in his activities whenever he thinks about the future.

The tendency to eliminate purpose and to concentrate instead on the description of mechanisms and and on the search for causes has progressively extended beyond the field of experimental science—where it had been adopted for pragmatic reasons—into other areas for which it was not originally intended, in particular the human condition.

If everything that happens is entirely the result of antecedent causes, it is senseless and indeed foolish to ask *why* things are thus and *what purpose* they serve. Once the idea of purpose is eliminated from the concept of nature, however, it becomes difficult to find any significance in the world. Even human life becomes literally meaning-less except for persons who have religious beliefs and regard man as somewhat apart from the rest of nature. The objective and purposeless philosophy of science has thus introduced into human life the feeling that the world is just what it is, without rhyme or reason.

A few philosophers have struggled to reintroduce meaning in the purposeless picture of the world that scientists have created, perhaps unwittingly, to facilitate their experimental investigations. For many people, however, what remains is simply the world as it is. Everything might have been quite different, and there would have been no reason for that either. The ultimate outcome of this attitude is the popular phrase of today: "Tell it like it is, man." Eventually the archetypal hero of the purposeless world might become Eichmann, the perfect functionary, who seems never to have experienced any human sentiment and simply did what he was told without concern for the meaning and ultimate consequences of his actions.

From what we can judge, the seventeenth century founders of scientific philosophy were pious men. They had no qualms about eliminating purpose from their studies of nature because they had no doubt about God's purposes. The pure rationality of the scientific method, as presently defined, proves inadequate, however, when applied to human life without consideration of purpose.

Utopians of the past tried to imagine or invent surroundings and ways of life conducive to happiness. Contemporary utopians, in contrast, are primarily interested in using science to manipulate external nature and man's nature. They postulate societies in which the Gross National Product will continuously increase; cheap nuclear power will enable man to squander energy; the reclamation of raw materials from junkyards and from the sea will permit man to continue being wasteful; foolproof contraceptive techniques and synthetic foodstuffs will eliminate the dangers of overpopulation; electronic robots will take care of the work in the kitchen and in the study room; means of communication will become ever more rapid through the air and through space; mechanical equipment will replace failing organs; drugs will control moods; and psychological conditioning and genetic manipulation will make it possible to design human beings according to specifications.

These imaginings of the future are not unreasonable; indeed, they are the products of rational thinking. They generate the nightmares of reason only because modern utopians deal only with means and are not concerned with purpose. Purpose must be introduced into rational thinking if one wishes to judge how large a Gross National Income should be, how far it is safe to let technology interfere with man's direct perception of the world, and whether human beings should be conditioned psychologically or altered genetically.

The psychological and technological utopias that appear possible when described in the conventional terms of science and technology are undesirable and, in fact, unbelievable when considered from the human point of view.

Expansion for expansion's sake is objectionable because exponential curves grow to infinity only in mathematics, not in the real world. Societies, like organisms, are likely to perish when their size or complexity reaches a point that creates demands which the environment cannot satisfy. As H. G. Wells remarked in "Mind at the End of its Tether," "In the records of the rocks, it is always the gigantic individuals who appear at the end of each chapter." Scientists and technologists must now concern themselves not so much with growth as with the control and direction of the social enterprise. Our societies have entered a phase in which control and direction of technology is more important than its expansion.

There is danger also in letting machines govern our lives for the sake of efficiency. By "playing the machine game according to *its* rule, man is almost bound to become a robot, although he may be a joyful one who will spend less and less time mourning his lost freedom." (Wilkinson, 1967)

It is not even certain that man can survive in the supersonic and supersensory world that certain utopians advocate; the environment would probably be incompatible with certain unchangeable aspects of man's nature. The man of flesh and bone can maintain physical and mental sanity only to the extent to which he can have direct contact with a certain kind of reality not very different from the conditions under which he evolved. Furthermore, he experiences happiness only if given the chance for spontaneous expressions of his whole personality. "Reason," wrote Dostoevsky, in the *Letters from the Underworld,*' "can only satisfy the reasoning ability of man, whereas volition is a manifestation of the whole of life . . . . Reason knows only what it has succeded in getting to know . . . whereas human nature acts as a whole, with everything that is in it, consciously, and unconsciously, and though it may commit all sorts of absurdities, it persists."

Artists and writers, who generally perceive and express the mood of the time before the general public has crystallized its own feelings, have stated in many forms since the beginning of this century that they are not happy or content with the modern world despite its technological triumphs and the comforts it provides. The disenchantment is now affecting many people other than artists and writers, especially the younger generations.

Two centuries ago young adults stormed the Bastille and other symbolic representations of political despotism. Today countless young men and women all over the Western world publicly or privately reject scientific knowledge, efficient technology, the ever-expanding economy, the search for happiness through greater affluence, and other values of the society of consumption that had been regarded as the very expression of desirable modernity by their forebears.

It is not sufficient, however, to protest against the past; nor is it possible to destroy its influence. In England, a famous theologian advocated a moratorium on science because he felt that scientific knowledge was growing faster than it could be assimilated and safely utilized. But this is idle talk. The theoretical and practical knowledge which made possible modern technology and the nuclear bombs is now part of the collective memory of mankind. Furthermore, nothing short of catastrophic upheavals can extinguish the human will to enlarge the body of knowledge on which scientific technology is based. No moratorium on science can still in curious minds the eagerness to learn more of man's nature and his relation to the universe. It is futile to burn books.

Visions and Disenchantment

There is no way for mankind to retreat from reason or from science; but this does not mean that mankind must continue on the road it is now following. *Trend is not destiny.* The world of tomorrow will express the image that man is now creating out of science and technology. At present the quality and characteristics of human life are determined more by accidental technological imperatives than governed by chosen values; they are more foisted on man than willed by him. The present century is called the technological age, not because there is a great abundance of machines and man is dependent on them but because we accept the fact that our lives are the manifestations of consequences rather than the expressions of purposes.

Despite our scientific and technological triumphs we suffer from a loss of nerve and have become a conservative society satisfied with continuing on our present course. We are no longer willing to construct models of possible futures that we really desire, despite the fact that our willingness to let science and technology proceed on their own course generates the nightmares of reason. Purpose naturally involves the willingness to take risks, but as Alfred North Whitehead wrote, "Adventure or Decadence are the only choices offered to mankind. The pure conservative is facing against the essence of the Universe." Scientists must use logic in their day-to-day activities, but they must be willing to make choices and take new departures if science is to become a really creative force in the social evolution of mankind.

Science at the World's Fair

We are witnessing great and irreversible changes not only in our ways of life but also in our views of the cosmos and of man's place in the order of things. Modern science has certainly played a large part in these changes through the application of technological and medical knowledge and the breakdown of traditional beliefs and attitudes. However, the extent of our present triumphs and ordeals does not warrant the common belief that modern science has been the only, or even the greatest, revolutionary force in human history.

Now, as in the past, men derive the impetus and moral energy for historical movements from passions and urges that have not changed since Homer personified them as the real heroes of the *Iliad* and *Odyssey*. All attempts to change the world of men and of things are still influenced by the ancient dreams and myths of mankind, as well as by the desire for rationality that has generated countless religions, social philosophies, and political systems.

Le Roy's book *Vicissitude,* mentioned in the first chapter of this book, illustrates that many aspects of sixteenth century life in Europe had been revolutionized by the practical inventions and advances in knowledge that occurred during the Renaissance, long before experimental science began to influence technology or act as an effective social force. Inventions and new knowledge, in fact, had caused social upheavals long before the Renaissance.

Visions and Disenchantment

At the risk of appearing paradoxical and being accused of *lèse majesté*, I must state my conviction that human life has been affected more profoundly by prehistoric experiences than by modern science. Humanness has been shaped by these experiences. The use of fire, the domestication of plants and animals, the development of agricultural and irrigation techniques, the emergence of cities, and the creation of religious doctrines and political systems are but a few of the technical and social innovations of the distant past that have given human life some of its most important characteristics and have determined how we respond to the world of today.

The fundamental changes in the ways of life and in the human environment that have occurred throughout history have not resulted from the applications of theoretical knowledge. The early Middle Ages was a period of great technological change in Western Europe. The heavy-wheeled plough first turned the sod of England at that time; improvements in the design of the wind mill, the water mill, and the harnessing of horses greatly increased the supply of power. The commercial exploitation of technology in medieval Europe was comparable in its pervasive effects to that of the Industrial Revolution some 500 years later. (Kranzberg 1968; Nef 1961; Forbes 1963).

Many techniques which constitute the bedrock of today's medical and surgical practice were developed without benefit of theory yet had effects on health far greater than those of twentieth century scientific medicine. This is the case for the most useful forms of vaccination (such as against smallpox) and for the practices of sanitation introduced during the nineteenth century. Also, a large number of drugs that are still in use today such as opium, quinine, digitalis, reserpine, and salicylate were discovered and first used empirically long ago by practitioners who were ignorant of medical science.

Other inventions such as the various forms of clocks, the printing press, the steam engine, and countless others were first put to human use on a large scale by practical technologists who were little more than artisans and had no knowledge of theoretical science.

The first phase of the Industrial Revolution began in England around 1600 when attempts were made to convert iron ore into cast and bar iron using coal as fuel, and steam was utilized as a source of power for the machinery used in draining coal mines. By the mid-eighteenth century Europe could boast of many steam–pumping engines and charcoal–consuming blast furnaces. Tracks of parallel rails fastened to the ground were first used for horse–drawn wagons and later for locomotives drawn by steam power. These early horse–drawn railways eventually led to the railroad age, much as industrial use of burning coal led to the age of iron and steel and the draining of coal mines led to the age of steam.

The art of making steel became progressively more efficient through empirical procedures; similarly the automobile industry and even aviation were first developed during the nineteenth century by talented tinkerers. In all these technological fields the inventors worked by trial and error and were unconcerned with the theoretical knowledge underlying their discoveries and inventions.

One does not have to travel far in space or time to gain a view of inventors who had no place in the scientific establishment yet developed techniques that have revolutionized life and indirectly have affected the course of science. In the Ford Museum in Dearborn, Michigan, there is a painting of Henry Ford as a young man working in an improvised laboratory; this laboratory was the kitchen of his rented flat in Detroit. The painting shows Henry and his research assistant (his wife) experimenting with an early combustion engine; Mrs. Ford is carefully feeding gasoline into a chamber sparked by an extension cord, while Henry is cranking the piston. During approximately the same period Edison was working in his private laboratory in New Jersey, as illustrated in photographs and paintings at the Edison Laboratory National Monument. Both Henry Ford and Edison were completely untrained in science and did not have any interest in theoretical inquiries. According to his own account, Edison selected problems for which there were demands from the industrial markets around him. Yet who would doubt that through their systematic empirical studies Ford and Edison have changed the ways of life all over the world.

Even though factories are now considered the mark of modern scientific technology, they came into existence long before the advent of modern theoretical science. The word "manufacture" means literally "making by hand." In the past there were factories that employed many workers but had no machines. For example, Jack of Newbury, who was the most famous clothier of sixteenth century England, employed 200 workers who tended 200 looms in one single large room, 100 women who carded the wool in another room, and 40 men who worked in a dyehouse (Nef, 1961). Such grouping of operations provided a social structure which was as important for the success of the Industrial Revolution as was the development during the eighteenth and nineteenth centuries of the mechanical loom, the spinning jenny, the steam engine, and the railway.

Thus, technology emerged and reached a high level of development without the help of theoretical scientific knowledge. In a speech delivered at Yale University in 1966 Professor Frederick Seitz, who was then President of the National Academy of Sciences, showed that technology could have continued developing without the help of science. Since this possibility is rarely mentioned it seems worthwhile to quote President Seitz at some length:

> Where would technology be today if the scientific method had not been developed? What would have happened if the Western Europeans had been somewhat like the Romans and devoted their talents almost exclusively to the everyday practical affairs of men, pushing ahead along traditional lines, much as the Romans pursued military and civil engineering without devoting

very much attention to the philosophical ideas of the Greeks? . . .

. . . matters up to 1800 probably would have continued more or less as they did if science had not developed or had remained isolated from technology. Most of the products of classical ceramics and metallurgy would have come in to use; coal and steam would have been exploited for heat and power. The age of geographic exploration would have gone on almost unchanged and stimulated the development of nautical instruments including the chronometer. The age of petroleum would have blossomed much as it did in the middle of the last century and we would have had the automobile and mechanized farming equipment. Clearly, the internal combustion engine would have taken on a different aspect because of the absence of electrical ignition, but we would still have had diesel-like engines and gas turbines, perhaps using hypergolic fuels or other means of ignition.

On the side of the life sciences, plant and animal breeding would have continued to progress along traditional lines; medicine would have evolved, probably with the discovery of more elementary anesthetics. The life sciences, however, as in past centuries, would have been clouded with mysticism. Quasi-mystical approaches to the technology of living systems resembling Lysenkoism would have been a common phenomenon everywhere. . . .

. . . eye glasses, the telescope, and the microscope would have come along in at least rudimentary practical form following the routes of traditional technology. After all, the spectacle and the telescope were devised well before optics became much of a science. Kepler was the first to give what might be called a modern theory for the telescope–however, it was proffered after the invention.

The first really significant omission which would have had a major influence upon technology would have been the science of chemistry. . . . In brief, many aspects of modern living would resemble somewhat those familiar in Europe and the United States in about 1875, with the one great exception that there would have been a continued evolution of sophisticated mechanical gear, such as the gasoline engine. Since farm machinery, the automobile, and possibly the airplane, would have emerged as a result, urban society would have continued to grow. Transportation would have been at least as rapid as it was in the first half of this century, even though the standard of living in the Atlantic community would have been somewhat more primitive. I think it seems safe to say, however, that without the evolution of chemistry the technologist would have been hitting his ceiling by the present time, because the materials available to him would not have evolved nearly as rapidly as his needs.

The second great omission we would have experienced, almost as serious as that we would have faced without the science of chemistry, would have been the science of electromagnetic phenomena. (See also similar statements by I. I. Rabi (1965) on page 127.)

While science had toiled almost obscurely in the rear of empirical procedures for a very long time, it began to step forward as the leader carrying the torch at the end of the past century. This change occurred progressively as illustrated by the use of steam power. The primitive steam engine was invented by Newcomen in 1705 and was much improved by James

Watt in 1765. When the use of the Watt engine became widespread the need for evaluating the yield of energy per unit of fuel consumed, as a basis for improving the efficiency of the machine, led the young theoretical physicist Sadi Carnot to investigate the quantitive relation of heat to power. Study of this relation continued to occupy the minds of physicists for almost a century. Joule, Meyer, Kelvin, and Helmholtz finally supplied the theoretical information from which the modern world learned to harness steam power for transportation and industry. Railroads, steamships, and the power plants of large factories soon emerged from the calculations and experiments of theoretical scientists whose studies had their origin in the empirical use of steam power.

The passage of electricity from the cabinet of the natural philosopher into workshops and homes appeared as a great miracle to the man of the nineteenth century; it convinced him that Bacon was right in stating that science is power. In 1819 Oersted found that an electric current tended to twist a magnetic pole around it. Shortly thereafter, the theory of the interaction between currents and magnets was developed by Ampère, who also pointed out that the deflection of magnets by currents could be used for telegraphic transmission. It was not long before Morse and Wheatstone had made a practical reality of the electric telegraph.

In 1823 Faraday showed that a wire carrying a current could be made to rotate around the pole of a magnet; he thus created the first electric motor. The electromagnet and the commutator were invented by Sturgeon during the next few years, and about 1830 the work of Joseph Henry in America and of Faraday in England led to the discovery of electromagnetic induction. The scientific armamentarium which made possible the dynamo and other electromagnetic machines was thus complete.

Although the practical achievements of science during the early nineteenth century were most spectacular in the production and distribution of power, other kinds of scientific pursuit also helped introduce science into everyday life.

Lavoisier is best known for having initiated the modern era in theoretical chemistry. Throughout his life, however, he was also deeply involved in the management of national industries in France. For example, he improved the manufacture of saltpeter and gunpowder and his classical studies on the composition of air originated from his efforts to design better lanterns for the lighting of Paris.

Pasteur, who began his professional life in physical chemistry, early became involved in the development of practical methods for the production of alcohol from the sugar beet; from there he moved to the scientific problems posed by the manufacture of beer and wine, and of silk from silk worms. These practical problems provided him with the inspiration for the profound theoretical studies that eventually revolutionized fermentation industries and the practice of medicine. (Dubos, 1950)

In England the mine explosion of 1812 near

Gateshead-on-Tyne led Humphrey Davy to study fire-damp and to demonstrate that the danger of explosion could be minimized by fine gauze put over the miner's lamp. His invention of the safety lamp in 1816 decreased the hazards of coal mining and thus contributed to the industrial supremacy of England. James Clerk Maxwell's highly theoretical studies of waves were published a few decades later. They were directed to specialists and had no meaning for the general public until they led to wireless telegraphy.

In Germany Wohler's synthesis of urea in 1828 opened the way for the synthesis of medicaments and dyestuffs. A little later Justus von Liebieg organized the first laboratory of biochemistry in Giessen. Stimulated by the desire to improve farmland he undertook studies which elucidated the principles of soil fertility; this lead to the rational use of fertilizers and thus provided a scientific basis for modern agriculture.

These examples, selected almost at random from nineteenth century scientific and social events, illustrate the nature of the changes that took place at that time in the relationship between science and technology. Scientists did not displace inventors. What occurred instead was the spontaneous development of social mechanisms which favored the constant feedback between theory and practice and made the distinction between the two increasingly tenuous.

Once begun, the interplay between science and technology went on at an accelerated rate. There is no need to illustrate that all the great inventions derived from physics and chemistry in our own century would have been impossible without the knowledge acquired in theoretical laboratories. And, as every scientist knows, much of modern science is now dependent on the esoteric products and specialized equipment produced by technology. The theoretical and applied aspects of science are now inseparable.

Increasingly those aspects of life which are called modern are carried out with tools derived from theoretical science, and more importantly the very characteristics of human life and of science are being shaped by scientific technology.

In 1899 A. R. Wallace, who formulated the theory of evolution simultaneously with Charles Darwin, published a book entitled *The Wonderful Century* in which he presented an enthusiastic account of the achievements of his age. He credited twenty-four fundamental advances to the nineteenth century, as against fifteen for all the rest of recorded history. Granted that these claims were exaggerated, Wallace's intent was nevertheless correct because nineteenth century science had done more than change the practical aspects of human life. It had revolutionized man's view of the cosmos and of himself.

The scientific geniuses of earlier centuries had, of course, profoundly influenced philosophy and human attitudes, but this influence had reached only a small percentage of the population. Only the intellectual elite and the dilettantes were affected by Copernicus, Galileo, Newton, or Laplace. In contrast, Darwinism rapidly spread through the general public, and, as a result, science began for the first time to live up to Pope's admonition that the proper study of mankind is man.

Then as now the public was interested not so much in science per se as in its relevance to human life. The theory of evolution was of extreme importance because it had a direct bearing not only on man's origin and on his place in the order of things but also on his religious beliefs and his attitudes toward economic, political, and social institutions.

The rate of acquisition of new knowledge was so slow in the past that ancient civilizations found it difficult to conceive of the possibility of progress. Men whose lives depended entirely on the course of natural events were more aware of the recurrence of daily and seasonal phenomena than of the continuous process of change which we now take for granted. Seeing that natural events repeat themselves year after year, they tended to extrapolate from these cosmic cycles to human history. For them the myth of eternal return seemed to apply to the affairs of man just as it did to the cycles of nature and the motions of stars. The known conditions of the present appeared as but one stage in the endless ebb and flow of events.

There is no way to determine with precision at what time the myth of eternal return was displaced in the Western mind by the concept of progress. The philosophical speculations of the Renaissance and of the Enlightenment certainly helped to formulate it and to make it intellectually acceptable. In any case the belief in a continuous process of change toward a new and better state different not only from the present but also from anything in the past became part of collective consciousness in the Western world approximately at the time when experimental science first began to prosper. Men like Condorcet or Franklin wrote of progress as a theoretical possibility, and they placed their hopes for its realization in the future developments of science. The doctrine of biological evolution provided the theoretical basis for their belief in progressive historical change.

Few laymen have an exact understanding of the scientific mechanisms involved in biological evolution. Yet most of them now accept as a fact that everything in the cosmos, heavenly bodies as well as living organisms, has developed and continues to develop through a process of historical change. In the Western world most theologians also accept a progressive historical view of creation.

Evolutionary concepts are widely and tacitly applied not only to living organisms and to man but also to his institutions, his customs, and his arts. These concepts appear so obvious that most religions, political parties, and schools of sociology, history, or art now make them the basis of their doctrines. It can be said without exaggeration that theoretical biology has thus introduced into human thought a new element which pervades all aspects of traditional culture.

Cosmology and the physicochemical sciences are also beginning to influence lay thought. As a biologist ignorant of these fields, I here represent the lay public whose views concerning the cosmos and the structure of matter are progressively being transformed by a kind of knowledge that does not involve real understanding. Like every human being I have been puzzled, for example, by the concept of the divisibility of matter. Nevertheless, I can comprehend that when sufficient energy is applied to elementary particles in the accelerators, the particles are changed, not by a process of true division, but by a transmutation of energy into matter.

I have just mentioned phenomena completely foreign to my personal knowledge in order to illustrate the manner in which science becomes integrated into the cultural tradition. Science shapes culture, not only through its technical aspects, but also by providing new points of view and generating new attitudes. That the earth is round and that all known living creatures have a common ancestry is not obvious either to my perception or my common sense, yet these concepts have become integrated into my daily thoughts and thus constitute part of the fabric of my culture. It would be surprising if the general concepts of particle physics and of relativity theory did not in some way affect the thought processes of the next generation.

The most concrete and spectacular evidence of the euphoria about science during the late nineteenth and early twentieth centuries was the popular success of the great international expositions (world's fairs) that were organized during this period. The attendance figures for a few of them will suffice to give a quantitative measure of the public interest: Paris (1889), 25,000,000; Chicago (1893), 28,000,000;

Paris (1900), 47,000,000; St. Louis (1904), 24,000,-000. These figures, large as they are, do not indicate the intensity of the excitement among intellectuals and the populace caused by the scientific and industrial exhibits.

It is probable that the organizers of the expositions were concerned primarily with the enhancement of national prestige, the display of industrial and commercial showpieces, and of course financial success. But the expositions also glamorized the transformation of human life by technology and generated hope for a scientific utopia. For example, the general public was exhilarated by the huge rotary printing press first introduced at the International Exhibition of 1878 and by the steam engines and dynamos that became the show pieces of later exhibitions.

In his *Education,* published posthumously in 1907, Henry Adams devoted a chapter entitled "The Dynamo and the Virgin" to the Great Exposition of 1900 in Paris; he tried to visualize how man would be affected, especially in his spiritual life, by theoretical and technological innovations. There was no doubt in Henry Adams' mind that faith in the power of religion was being replaced in all social groups by faith in steam and electric power. From worship of the Virgin Mary medieval man had derived the inspiration and also the energy to build his civilization. In contrast, modern man was now dependent on scientific technology not only for the construction of machines and buildings but also for giving direction to his social and emotional desires.

The dynamo, along with the steam engine, was the most popular of the exhibits at the Exposition. To Adams it became a "symbol of infinity." As he grew accustomed to the great gallery of machines he began to "feel the dynamos as a moral force, much as the early Christians felt at the Cross."

Beyond displays of power there was something else in the Exposition of 1900 that greatly perplexed Henry Adams. For the first time X-rays and radium were being introduced to the public, and their mysterious character revealed an entirely new order of things as foreign to orthodox physics as to human experience. The physicist Langley, who acted as Adams' guide, accused these new forces of being anarchical and of denying the truths of the science he knew.

Adams realized that the science of 1900 was "not the first to upset school masters. Copernicus and Galileo had broken many professorial necks about 1600; Columbus had stood the world on its head towards 1500; but the nearest approach to the revolution of 1900 was that of 310, when Constantine set up the Cross."

The newly discovered X-rays were "occult, supersensual, irrational; they were a revelation of mysterious energy like that of the Cross." Adams could find sense in electric and steam power, frozen air, or the electric furnace, "but X-rays had played no part whatever in man's consciousness, and the atom itself had figured only as a fiction of thought. In these seven years man had translated himself into a new universe which had no common scale of measurement with the old. He had entered a supersensual world, in which he could measure nothing except by chance collisions of movements imperceptible to his senses."

Henry Adams' uneasiness as he faced the dynamo, X-rays, and radium seems now as outdated as his style. A measure of the extent to which modern man takes for granted the most esoteric and powerful manifestations of science is the general nonchalance of the public before the scientific exhibits at the recent international expositions.

The 1963 World's Fair in Seattle, for example, glorified two different but equally spectacular aspects of modern science: the production of nuclear energy and the biochemical mechanisms of life. Among the showpieces of the Fair, still on exhibit at the Pacific Science Center in Seattle, were the multicolored panels used for operation control in nuclear energy plants and the giant, multicolored model of DNA, the complex molecule which is touted to contain the secret of life. Had he visited the Seattle World's Fair Henry Adams probably would have contrasted the lofty and esoteric atmosphere created by these colorful exhibits with the reverential and mystic atmosphere of the stained glass windows in the gothic cathedrals. But the expressions "nuclear energy" and "DNA" now share newspaper headlines with reports of gangsters, drug addiction, and sex scandals. All reverence is gone.

Few people seem to mind, as Henry Adams did, that the view of matter and of life as conceptualized by science have become increasingly and perhaps irreversibly meaningless except for specialists. At the international expositions man is now claimed to be a machinery of atoms, complex of course, but nevertheless explainable in ordinary chemical terms; human life is depicted as a physicochemical operation that will increasingly be powered and governed, so it is claimed, by nuclear energy. Unamuno's man of flesh and bone (1954) seems almost irrelevant in the world of machines, and there is no place for him in the scientific exhibits of the up-to-date world's fairs.

Canada tried to go beyond the machine in Montreal's Expo '67 which was organized around the theme "Man and his World." Despite the phraseology, however, none of the exhibitors had much of importance or even of interest to show or say concerning man himself. The U.S. geodesic dome designed by Buckminster Fuller was startling, but it chiefly housed equipment for space travel and for the trivialities of mechanized daily life. The Russian Pavilion derived its human timeliness from its profusion of political propaganda. A more suitable name for Expo '67 might have been "The World of Human Machines and of Human Conditioning." As these pages are being written San Antonio, Texas is holding its exposition named "Hemisfair," which emphasizes man's flight into space but betrays, in reality, his alienation and flight from the earth and from his biological reality.

The world's fairs have illustrated the achievements of science, but they have also displayed the illusions that are being fostered by unwarranted promissory statements concerning the potentialities of science.

The following quotation taken from an article published in 1967 shows that sophisticated scholars are even more gullible than the general public regarding scientific potential: " 'Today we feel that there are no inherent secrets in the universe . . . and this is one of the significant changes in the modern moral temper.' For the first time since the Greeks, man is convinced of the essential intelligibility of the universe: there is nothing in it that is in principle not knowable." (Mesthene, 1967)

The naiveté of the faith in the explanatory power of science displayed in the previous quotation would be disturbing if it did not provide us with perspective because of its similarity with a statement made by Marcellin Berthellot during the second half of the nineteenth century: "We are entitled to believe . . . that we can create anew all the substances and creatures that have emerged since the beginning of things . . . . The world has no longer any mystery for us." This was written by one of the most eminent chemists of his age, in 1885!

A bulky and highly entertaining anthology could be made of the misleading statements and hopes concerning scientific and technologic innovations that have been made not only by laymen otherwise famous for their wisdom but also by scientists and technologists. Two examples will suffice to warn against the kind of optimism that is frequently generated by spectacular new technologies.

In a sermon given in 1621, John Donne advocated the view that more deadly weapons might be turned to human benefit: "They have found out *Artillery* by which warres come to quicker ends than heretofore, and the great expense of bloud is avoyded: for the numbers of men slain now, since the invention of Artillery, are much lesse than before, when the sword was the executioner."

In the *Scientific American* for July, 1899, there was discussion of the probable effects that the motor car would exert on urban life once mass production had lowered its price: "The improvement in city conditions by the general adoption of the motor car can hardly be overestimated. Streets clean, dustless and odorless, with light rubber-tired vehicles moving swiftly and noiselessly over their smooth expanse, would eliminate a greater part of the nervousness, distraction, and strain of modern metropolitan life." If this passage were engraved in front of the steering wheel in our cars, every driver could meditate over human illusions as he sits motionless in the smog of a traffic jam!

Despite John Donne's hopes the development of *Artillery* has not decreased the destructiveness of war; optimists must now take heart from the hope that nuclear weapons will act as war deterrents because they are so much more destructive than conventional gunfire. Contrary to the assertions in the 1899 *Scientific American* article automobiles are poisoning the air and spoiling the quality of life with noise, confusion, and strains; but optimists may still hope that the senseless density of vehicular traffic will impose saner policies in the planning and management of urban areas.

Technological solutions have rarely solved social problems in the past, nor are they likely to do so in the future. They commonly provide an excuse for not facing these problems and thus constitute a form of escapism into gadgetry from the complexity of vital issues. Scientific knowledge on the other hand, can provide the understanding required for a rational approach to almost any kind of difficult situation. It can give policy makers the facts on which to base decisions and increase the numbers of options they have to work with. In the past the range of choices was sharply limited by physical constraints, but these can now be overcome by scientific knowledge and its derivative technologies. E. G. Mesthene (1967) has illustrated the consequences of the removal of physical constraints for decision-making in the field of international politics:

> The physical conditions of political action are no longer fixed. New devices to detect nuclear explosions were developed while arms-limitations negotiations were going on, and they affected the course and the results of those negotiations. Recent negotiations with Panama were clearly influenced by the earth-moving possibilities of atomic explosives. Both radar and the atom bomb were developed and deployed during and in time to affect World War II, and their development was undertaken deliberately for the purpose of affecting it.

It is unfortunate that the most conscious use of science and technology in the formulation of long-range policies so far has been in the domain of power politics. But this should not be cause for discouragement. The important fact is that inability to change the physical world is no longer an inviolable constraint in the shaping of social systems and institutions. The theme of a future world's fair might be not the display of scientific and technological stunts but the use of science and technology for the creation of surroundings and ways of life that are both possible and desirable.

4
Scientists and Society

Scientific Method and Social Choices

In 1962 an international symposium on "Man and His Future" was held in London. Although the presentations and discussions were informative and spirited they did not provide any basis for agreement as to desirable scientific goals for the welfare of man, let alone programs of action. At the end of the meeting one of the most eminent and influential participants, Sir Peter Medawar, expressed the view that the difficulty in achieving consensus about the scientific aspects of complex social issues was inherent in the very nature of science. His statement so well expresses the views held by the majority of scientists on the planning of science that it seems worthwhile to quote here a portion of his remarks:

Scientists and Society

> This is the thing that has impressed me most about this meeting—the sheer diversity of our opinions . . . . This diversity of opinion is both the cause and the justification of our being obliged to do good in minute particulars. It is the justification of what Karl Popper called "piecemeal social engineering." One thing we might agree upon is that all heroic solutions of social problems are thoroughly undesirable and that we should proceed in society as we do in science. In science, we do not leap from hilltop to hilltop, from triumph to triumph, or from discovery to discovery; we proceed by a process of *exploration* from which we sometimes learn to do better, and this is what we ought to do in social affairs. (Medawar, 1963)

If it were true that scientists always move step-by-step and are concerned primarily with "minute particulars" (a phrase borrowed from William Blake), then the most profitable and only safe way to use science in the practical affairs of man would be to proceed according to what Karl Popper has called "piecemeal social engineering." But the scientific enterprise is much more complex and socially intricate than that.

Scientists and technologists have to proceed piecemeal for each particular, limited item of theoretical study or of social application, but this kind of approach applies only to the *tactics of their daily work*; it is not applicable to the *strategy* of science. The kinds of "minute particulars" that scientists investigate and the "social engineering" in which technologists engage are largely determined by factors almost independent of science itself. Among these factors are man's conjectures and anticipations of the future. It is correct, as William Blake wrote, that "he who would do good in anything must do it in minute particulars." But it is equally true, again in Blake's words, that "what is now proved was once only imagined." (*The Marriage of Heaven and Hell, 1790*)

The importance of imagination in human life gives a special meaning to Aristotle's often quoted statement: "All men by nature desire to know." Each of us has his own views concerning what is worth knowing. Our search for knowledge is motivated and directed by an urge to create meaningful patterns out of the bewildering confusion perceived by our senses. We want to shape reality according to our own wishes and are so presumptuous as to personalize the Universe. Each of us lives, as it were, in a separate world of his own.

Our desires are infinite and extend into the distant future, whereas we are finite and mortal. What Unamuno called the tragic sense of life has its origin in our unwillingness to accept this finitude. We invent philosophies and scientific concepts as substitutes for the reality we cannot fully grasp.

At any given time the intellectual attitudes toward reality that are considered correct are those that are acceptable to the majority of people. This applies to scientific truth which can be regarded as a transient consensus. In order to achieve a position of social leadership science must not only develop procedures that facilitate daily life, but it must also provide interpretations that appear to give more explicit meaning to reality.

The bulk of scientific knowledge is acquired through the step-by-step accumulation of isolated facts, progressively organized into patterns or systems that have an internal logic.

Astronomers first examined the sky with naked eyes, then with light telescopes, then with radiotelescopes. They now take pictures of the heavenly bodies on emulsions sensitive to different kinds of rays; and they apply to the cosmos as a whole the laws of physics first discovered on earth. Even the layman can recognize a logical trend in this progressive development of techniques and knowledge.

The students of human evolution use scientific concepts and techniques that bear no relation to those used by astronomers, but they, too, use an approach that has a quality of logic. Step-by-step, they have traced the origins of man back to a small, shrew-like creature that lived at the beginning of the Pleistocene; they have identified the environmental conditions, ways of life, and evolutionary changes associated with the progressive transformations that made forest-dwelling primates evolve into *Homo sapiens;* and they are now beginning to concern themselves with the interplay between organic and social evolution. Here again is a logical sequence, but one which deals with life instead of inanimate matter.

Logical processes of thought can be readily recognized in all fields of knowledge. Furthermore, experience has shown that the findings in one scientific discipline are never incompatible with the findings in another discipline, even though the facts themselves differ profoundly in nature. Despite their disparities all natural sciences can thus be assumed to constitute a single, logical structure. There are indications that social sciences also will find their place in the general pattern of scientific knowledge.

Ever since the Renaissance experimental techniques and theoretical studies have yielded factual knowledge and abstract generalizations at an accelerated rate. The resulting spectacular growth of science is usually described as a cumulative, orderly process through which discoveries have evolved one from the other as by an inexorable logic. Such logical unfolding evokes a seemingly endless and rather dreary road marked at intervals by milestones indicating notable achievements.

The true history of science, however, is far more fallible and erratic than appears from the depersonalized account given in textbooks. This is true because the professional activities of scientists have always been profoundly influenced by the historical accidents and social forces that give such an unpredictable quality to all aspects of human behavior. Instead of being the dreary unfolding of an inevitable logic, science develops in a halting manner which appears at times irrational because scientists share the fashions, passions, anticipations, and illusions of their contemporaries.

Scientists and Society

The English geneticist C. D. Darlington compared scientific progress to the pulling of a drawer which gives on one side only to jam on the other (quoted in Butterfield, 1950); but this picturesque image does not adequately express the extent of the changes in direction that often mark the scientific effort. In many cases scientific revolutions follow the discovery of unexpected facts which cannot be reconciled with orthodox doctrines. In other cases the development of new techniques makes it possible to travel over scientific ground formerly impassable; as a result, unsuspected horizons are visualized. Whatever the reasons, the scientific community often behaves as if it had reached a crossroad. However, the factors which lead scientists to select one road among the many possible roads they could follow certainly play a large role in the advancement of science.

Some of the factors that affect the direction of the scientific effort are rather unexpected. It has been suggested, for example, that the practical problems posed by brewing practices greatly stimulated the early development of science in Europe. The process of converting barley into beer involved physicists in the problems of gas pressure, chemists in the structure of starch, enzymologists in the mechanisms of fermentation, and microbiologists in the study of yeasts and bacteria. Much has also been learned by psychologists about the effects of alcohol on man's behavior.

The social and economic atmosphere of competitive nineteenth century England provided an environment which was suitable for Darwin's views that natural selection accounted for biological evolution. The prevalence of class struggle and a misreading of the law of the jungle also had a share in making Pasteur, Koch, Metchnikoff, and other early exponents of the germ theory of disease regard the phenomena of infection as a kind of warfare between body and microbes—just as Marx and Freud made human conflicts the basis of their social and psychiatric doctrines.

Thus in any given period scientific activity embodies preoccupations and attitudes that are widespread in the community. This explains why so many discoveries have been made, and laws formulated, simultaneously by several scientists who knew nothing of each other's work (Merton, 1957, 1961). For example, the needs of the Industrial Revolution stimulated scientific interest in the production and utilization of power, hence the many simultaneous discoveries concerning the transformations of steam and electric power into other forms of energy.

Scientists generally feel most comfortable when they deal with problems that coincide with the preoccupations of their contemporaries. By acting as hewers of wood and drawers of water they help science move forward rapidly along a fairly predictable course. A few unusual scientists, however, perceive the interests and tasks of the future. They undertake pioneering work toward the solution of future problems and thus open new channels toward further scientific discernment. Even scientific innovations which appear to have occurred independently of social influences usually have roots in the preoccupations of the social milieu in which they arise. They are not aberrations in the sequence of cultural events, but rather they are the expressions of highly perceptive scientific mentalities through which social undercurrents become manifest. Scientific innovators may seem unrelated to their times, but in fact they are like prophets who give structure and voice to the shapeless aspirations of the multitudes among whom they live.

Science is more than the product of its internal logic acting on antecedent knowledge. Rapid changes in the direction of the scientific effort can be caused by the pressure of social forces and the influence of visionaries. This does not mean that all science can be traced to social motivation or to the search for practical applications. The pure scientist who continually pursues pure science creates a form of knowledge which has an intrinsic value and needs not be justified otherwise. But the great majority of scientists become consciously or unconsciously involved in public affairs and are motivated by preoccupations other than the search for truth.

Francis Bacon, who was the most articulate exponent of the new scientific faith of the seventeenth century, taught that knowledge was to be cultivated not for its own sake, but for the benefit of human life: "Knowledge, that tendeth but to satisfaction is but as a courtesan, which is for pleasure and not for fruit or generation . . . . The sure and lawful goal of the sciences is none other than this: that human life be endowed with new discoveries and power . . . . As in religion we were warned to show our faith by works, so in philosophy by the same rule the system should be judged of by its fruits, and pronounced frivolous if it be barren." By influencing the selection of scientific problems, the requirements of English economy in the seventeenth century played a dominant role in shaping one of the most brilliant eras of science anywhere in the world. Nor was this situation peculiar to England. In 1671, Louis XIV provided an endowment to the Paris Academy for experimental work and the dissemination of practical results. Throughout the eighteenth century, Paris was the capital of scientific thought, because France was the first country to put into practice on a large scale Bacon's precepts that scientific work must be organized, summarized, and propagated.

As mentioned earlier, many nineteenth century scientists also were influenced in the choice of their activities by the industrial and medical problems of their environment. And one would have to be blind to evidence not to recognize that practically all of contemporary science is an expression of social pressures—both praiseworthy and objectionable ones.

Scientists have been so effective in providing technical solutions to the problems posed by society that they have now achieved a position of unprecedented power and prestige. After the Second World War, and especially after Sputnik, they were given almost *carte blanche* with regard to the extent and direction of their activities. During the post-Sputnik era it was taken for granted that only research scientists were competent to determine what problems are worth studying and how the scientific enterprise should be managed. There was a widespread feeling that, as had been said for General Motors, what was good for science was good for the country.

A striking change in this laissez-faire attitude has recently become evident, especially during debates in Congress. Society no longer seems willing to entrust decisions concerning science to scientists; it wants to decide which scientific problems deserve highest priority in public support.

The rapid increase of public interest in the management of science derives in part from the soaring costs of scientific research and from the necessity to finance it by taxation. But there is also another reason which is more interesting. There has recently developed a widespread awareness that scientific advances are not only creating grave dangers for human life but also determining its trends and that therefore they must be subjected to social control. Few are those who advocate a moratorium on science, but the feeling is growing that the scientific enterprise should be made subservient to human ends.

During the ceremonies for the centennial observance of the National Academy of Sciences, President Kennedy stated that "scientists only can establish the objectives of their research, but society in extending support to science must take account of its own needs." (Kennedy, 1965) All over the world, legislatures and governments are now claiming the right to formulate the strategy of the scientific enterprise, leaving to scientists the responsibility to work out the tactics best suited to the attainment of social goals.

Many members of the scientific community, however, doubt that scientific research is amenable to social planning. Michael Polanyi, a scientist turned philosopher, has stated unambiguously that "the aspiration of guiding the progress of science into socially beneficent channels is impossible and non-sensical." John Maddox, who is the editor of the English journal *Nature,* believes that only specialists have the knowledge required for the planning and evaluation of science. According to him, "the essential need is the exercise of scientific judgment, similar in every way to the judgment in the decisions which must be made on the conduct of individual experiments." (1968)

In general, research workers are skeptical as to the possibility of planning scientific research toward social ends. Their skepticism derives in part from the ambiguity of their attitudes toward the relative merits of knowledge per se as well as its practical applications. Most scientists have nostalgia for the pristine pursuit and academic purity of their profession, but they nevertheless find it difficult to resist the intoxicating appeal of involvement in problems having social glamour. It is certain, in any case, that the scientific enterprise will increasingly fall under social control. One of the most obvious manifestations of this trend is the limitation imposed on the movement of scientists and their divulgence of scientific knowledge, especially in war time.

Scientists and Society

In 1813, at the height of the Napoleonic wars, the English government allowed the illustrious chemist Humphrey Davy to go to France because he wanted to study the volcanoes of Auvergne. Napoleon was equally as broadminded as the English and allowed Davy to travel, lecture, and experiment in France, and even collaborate with several members of the French Institute. Such liberality stands in sharp contrast to the attitude of American and European governments during the First and Second World Wars. It has been quoted as evidence that in the past "scientists were never at war" (Beer, 1965), intellectual interests formerly prevailed over political considerations. There is a less charitable explanation, however, for the tolerance exhibited by the British and French governments during the Napoleonic wars. It is simply that Davy's scientific knowledge of chemistry had no military significance.

The movements of scientists were restricted as early as the eighteenth and nineteenth centuries whenever practical matters were clearly involved. For example, in 1760 the Duc de Choiseul gave a passport to John Crauford of Auchenames allowing him to travel in France even though the Seven Years War was then in progress, but the passport carried the restriction that he should not pass through a French naval port; the state of the French Navy was a matter of national security. Also, when Baron Stein visited Cornwall in 1787, Boulton and Watt issued drastic orders to prevent him from inspecting the new steam engine which had just been installed under their direction; commercial interests prevailed over individual freedom.

Secrecy in matters of theoretical science and technology is now stricter than in the past because almost every aspect of scientific knowledge has potential relevance to national, economic, and other aspects of public life. In the words of the Russian physicist Kapitza, "Science has lost freedom [because] it has become a productive force." (Vucinich, 1968)

The increasing integration of scientific activities into the social order will make it necessary to constitutionalize science. (Harvey Wheeler, 1968) There was no point in legislating the activities of scientists until recent times, because scientific advances occurred infrequently and without conscious anticipation of their long-range social effects. But the situation is now different in two ways. One is that the systematic production of socially meaningful discoveries, first anticipated in Bacon's imaginary *New Atlantis,* has now become a reality. The other is that, contrary to what was believed until the last century, the advancement of knowledge does not automatically improve the human condition. The events of the past few decades have shown that it is dangerous for society to allow scientists to manage the consequences of their work.

Scientific planning naturally implies an interference with the free market in scientific ideas and pursuits. As in economics the laissez-faire approach to science is tolerated only as long as it does not produce results that are harmful to society. During the past decades nuclear warfare, the chemical pollution of air, water, and food, electronic surveillance, and the ecological crisis have constituted for the free market of science the equivalent of economic depressions for the production and exchange of goods and services. Public controls and regulations are as repellent to scientists as they are to businessmen, but they are just as inevitable. They will certainly take the form of attempts at minimizing the ill effects of science; more importantly perhaps, they should aim at enhancing its beneficial effects.

For lack of a more sophisticated scientific philosophy the first and simplest approach to the control of science will certainly be through the allocation of financial resources. Such an economic policy is, in any case, inevitable because the needs of the scientific community are increasing faster than the national budget. In several of the more advanced industrial countries science has come to consume two-to-three per cent of the Gross National Product; it is no longer possible to postpone the painful task of deciding how the limited financial resources should be allocated among competing fields, institutions, and individual investigators—all clamoring for increasing support. During the past few decades financial support for science grew so rapidly that most worthwhile projects could be financed. From now on, in contrast, scientists will have to agree on a system of priorities or to fight among themselves. The latter is the more likely alternative. Yet, it is unthinkable that the overall science strategy should be settled by default or dominated by the scientific groups which have the most effective public relations. (See H. Brooks, W. Carey, and A. Weinberg).

One of the most articulate analysts of science policy during recent years has been Alvin M. Weinberg, a physicist who has been research director for twenty years, then director of the Oak Ridge National Laboratory, and has repeatedly served on the most important committees concerned with the formulation of scientific policy in the United States. Weinberg has become increasingly concerned with the social applications of science, a concern which has led him, a physicist by training, to advocate greater emphasis on biological sciences. It seems worthwhile, therefore, to review briefly some of the thoughts he has expressed concerning the possibility of formulating "criteria of scientific choice" on the basis of scientific and social considerations.

From the purely administrative point of view it is necessary to make institutional choices as to the nature, size, and geographical location of universities or research institutes among which the available funds are to be allocated. Political as well as educational considerations will certainly play a large role in these choices.

From a more academic point of view other choices will have to be made on the basis of criteria internal to science and particularly on the basis of the quality of the scientific effort to be supported. For example, "Is the field ready for exploitation? Are the scientists involved in it of sufficient competence?"

Criteria internal to science are the ones most usually employed by committees of experts in allocating research grants. But, according to Weinberg, "It is not tenable to base our judgments entirely on internal criteria. . . . Science must seek its support from society on grounds other than that the science is carried out competently and that it is ready for exploitation; scientists cannot expect society to support science because scientists find it an enchanting diversion." (See also J. F. Kennedy, 1965)

Weinberg proposes three external criteria: technological merit, scientific merit, and social merit. As to the first criterion: "Once we have decided, one way or another, that a certain technological end is worthwhile, we must support the scientific research necessary to achieve that end. It is not always easy to decide the technological usefulness of a piece of basic research. However, it is my belief that technological bolts from the scientific blue are the exception, not the rule, and that solving a technological problem by waiting for spinoff from an entirely different field is rather overrated."

According to Weinberg, the other two criteria, scientific and social merit, are "much more difficult; scientific merit because we have given very little thought to defining it in the broadest sense; social merit because it is difficult to define the values of our society."

Scientific merit can perhaps best be defined "by proposing that, other things being equal, that field has the most scientific merit which contributes most heavily to and illuminates most brightly its neighboring scientific disciplines."

The evaluation of "social merit or relevance to human welfare and the values of man" is naturally the most difficult problem posed by the application of Weinberg's criteria. It is also the most important because, in the end, our values shape our knowledge, which in turn determines the quality of human life. Depending upon their professional experience, knowledge of human situations, and judgments of value, individual scientists differ as to the areas of science which they consider to have greatest social merit. It is fortunate, therefore, that the United States is a pluralistic society and can afford "to have a plurality both of criteria of choice and orders of merit." (Toulmin, 1968)

In a pluralistic world problems of scientific choice inevitably involve non-scientific considerations. Therefore, scientific choices will have to be based on an enlarged concept of scientific ethics. Traditionally, scientists have been satisfied with a professional ethic of means based on well-developed canons of intellectual integrity and open-mindedness. They have not been much concerned so far with an ethic of ends involving canons of goodness, justice, and beauty applied to the applications of science. (Wheeler, 1968) From the social point of view, however, the ethic of ends is at least as important as the ethic of means. An entirely different attitude is required to deal with a situation to be explained (by theoretical science) and a function to be performed (by applied science).

The first is ethically neutral, whereas the second is intrinsically ethical. The classical professions (especially medicine, law, and divinity) grew from functions to be performed, and therefore had to develop complex codes of ethics derived from their social roles.

Since all fields of science now generate devices and services that increasingly affect human life, scientists must develop ethics of ends in addition to their traditional ethics of means. Each field must possess not only *techne*—its technique—but also *episteme*—its philosophy. The present preoccupations of governmental bodies and of the general public with the strategy of science will demand that scientists formulate the social philosophy of their respective disciplines with more care. This is particularly urgent for the environmental sciences, since all the manipulations of the environment will profoundly affect the biological and mental well being of man now and in the future.

The public demand that scientists become involved in environmental problems is illustrated by the following statement from the "Report of the Subcommittee on Science, Research and Development" (H.R. 89th Congress): "Problems of living in today's environment are reaching proportions which are truly monumental; they will not be solved without an equally monumental lift from science and technology." The same report also expresses the feeling that the scientific establishment is not properly organized to study the problems of modern life. For this reason it urges that the National Science Foundation be reorganized to better keep abreast of the nation's scientific needs, especially those related to environmental problems which have already reached a critical stage and those likely to emerge in the near future.

Technologies to manipulate the environment are becoming more numerous and more powerful each day, but they will not contribute to the improvement of human life until they are applied under the guidance of an environmental philosophy based on an ethic of ends.

It would be surprising if the concern of the public and governmental agencies for the environmental crisis and consequently for the ethics of scientific technology did not bring about a profound change in the direction of science. When it occurs, this change will result not from a logical process inherent in scientific knowledge and the scientific method but rather from deliberate social choices based on a judgment of values.

On Tap, But Not on Top

Scientists have been accused of leaving their discoveries like foundlings on the doorstep of society, ignoring the foster parents who do not know how to bring them up. This accusation, however, is no longer quite as justified as it once was. There are socially minded scientists who are as concerned with the utilization of knowledge as with its production; some of them even subscribe to Thomas Jefferson's dictum that "The end of all knowledge is action." On the whole, however, knowledge for knowledge's sake remains the professed ideal of the scientific community.

Until a few decades ago the most effective approach to the betterment of human life was to discover the workings of natural phenomena, because the theoretical knowledge thus acquired could usually be converted into practical applications. The important problem was to learn *how* to do things, because there was so little that could be done in practice. This attitude is still valid but no longer sufficient. Experience has shown that technology based on the experimental method makes it possible to create many more things than can be usefully or even safely utilized. Therefore the main problem is *what* to do among all the things that can and should be done.

As mentioned earlier, in most Western countries, the public is no longer willing to let scientists make policy decisions concerning the social use of knowledge. This does not mean that our societies are rejecting science, or even that the anti-science movement has spread beyond limited fringe groups. It means, however, that many enlightened persons agree with the English statesman who recently stated that scientists should be on tap, but not on top. This view of the relationship between science and the social order is particularly widespread among members of Congress and in government agencies.

Western culture has always been strangely ambivalent toward science and scientists. Admiration of scientific achievements is mixed with awe, yet the public is not stimulated to understanding the processes of science. Despite countless exhortations few consider science an interesting intellectual discipline or scientists (qua scientists) stimulating members of the intellectual community. The attitude of modern societies toward the scientist has much in common with the attitude of primitive people toward the shaman—he is a member of the group regarded as essential to tribal life, yet he is feared and at times hated. Throughout Western societies and even on college campuses, science is like a foreign power which does not represent either the feelings or the aspirations of the collective soul.

The alienation of science from society may be due partly to the fact that the problems which are most peculiarly human are not yet readily amenable to scientific inquiry. In any case, these problems are grossly neglected by scientists. The feeling of alienation has recently been increased by the fear that science is leading man into adventures that he does not really want. In his novel *The Machine Stops*, E. M. Forster describes scientific technology as an independent force "that moves on, but not on our lines; that proceeds, but not to our goals." As we have seen, novelists, poets, and science fiction writers tend to present science as a force that man created but that he can no longer control. Even more striking is the use of the theme of the sorcerer's apprentice by several famous scientists such as Max Born (1968) or Norbert Wiener (1950, 1966), and by sociologists such as Jacques Ellul (1964; Stover, 1963) and J. K. Galbraith (1967). In *The Technological Society*, Ellul describes as a *fait ac-*

compli the take-over of society by anonymous technological forces which operate independently of human control. In *The New Industrial State,* J. K. Galbraith asserts that the modern technological society is an almost self-contained system responsive only to the direction of an essentially autonomous "technostructure." Although the system still depends on public acceptance, it secures the acceptance of its products through an artificial demand created by advertising and government policies. Of course man has not become a completely passive creature of scientific technology, but it is painfully evident that he finds it increasingly difficult to free his life from the technostructure.

The scientific community is progressively becoming more aware of the dangers inherent in the social misuse of its creations. In countless conferences, seminars, and speeches scientists acknowledge their duty to evaluate and control the social consequences of their professional activities. But this expression of social concern rarely leads to corrective action.

Most scientists continue to select problems on the basis of intellectual interest and especially of prevailing fashion, rather than of importance for human welfare. There are good reasons for this attitude. One reason is that many scientists are interested primarily in theoretical and abstract issues. This kind of dedication accounts in large part for the success of the scientific enterprise as an intellectual pursuit. Another reason is that it is much easier to deal effectively with theoretical scientific problems than to formulate desirable and attainable social goals. As Einstein is reputed to have said, sociology is more difficult than physics.

Many scholars give lip service to practical problems in order to pacify their social conscience. However few are willing or find it possible to make any significant change in their scientific activities. A few examples of this intellectual escapism will explain why so little is being done to focus scientific effort on urgent social problems despite so much talk on the role of science in human affairs.

A common form of escapism is exemplified in the assertion that, in the long run, all scientific achievements contribute to the welfare of mankind. Scientists can quote much historical evidence to support the view that even when their work seems unrelated to the problems of daily life, it nonetheless serves the common cause by helping the public make intelligent use of available knowledge.

Furthermore, throughout the maelstrom of the scientific and technologic enterprise the almost mystical belief prevails that every innovation will eventually be of use to mankind. Every physicist believes subconsciously that new insight into the constitution of matter will result ultimately in a piece of useful hardware; every chemist and engineer expects that the new products, gadgets, or services that he develops for the market will in some way improve health, make life more comfortable, and increase both productivity and leisure time; every theoretical biologist entertains the illusion that his work will advance the control of cancer, heart disease, or other ailments of the modern world.

Like other human beings, scientists are bewildered by the accumulation of goods that science makes possible, and they are aware of the dangers inherent in material prosperity—from anomie to environmental pollution. Most of them, however, take it for granted that science eventually will provide solutions for all technologic and ecologic problems. The public naturally tends to welcome any expression of hope for a foolproof approach to modern life. This form of collective escapism is seen in Ortega y Gasset's "mass man," who feels no urgency about the ecologic crisis, overpopulation, or racial violence because he has heard that technologic solutions will rapidly provide temporary relief for these problems and thus give time for more permanent solutions at some unspecified time in the future.

The recourse to "technologic fixes" constitutes an escape from the necessity to face up to painful social problems. Similarly, according to Norbert Wiener, the use of automated decision-making equipment is motivated "in part by the desire to avoid direct responsibility. I am sure that a lot of the subdivision of effort in secret projects and highly compartmented projects has the same motive. The subordinate does not know enough about the project to feel responsibility and the man in charge can place responsibility with the system. I believe that one of the greatest dangers at the present time has to do with the attempt to avoid responsibility in order to avoid the feeling of guilt."

The simplest form of escapism is to believe that committee work is an effective approach to the solution of social problems. The fallacy here is not so much that, as the wag has it, committees keep minutes and waste hours; rather it is that the interest of committee members in the problems assigned to them is at best transient and usually superficial. As a result they hardly ever do anything to implement their own recommendations. Many current examples could be quoted to illustrate this statement, but for the sake of propriety, it seems wiser to go back a few years in time.

Shortly after the publication of Rachel Carson's book *Silent Spring,* President Kennedy appointed a blue ribbon committee to advise him on the pesticide problem. All the members of this committee were eminent biologists who held influential faculty posts in great universities. Their report confirmed facts that had been long known by anyone familiar with the field, and it also emphasized the need for further study. Despite the sentiments expressed in their report, however, there is no indication that any member of the committee did anything to encourage his staff or graduate students to work on pesticide toxicity. Nor is there any evidence that the committee has fostered such research in other institutions during the several years that have elapsed since the preparation of the report.

The tendency now is to make blue ribbon committees even more impressive by referring to them as task forces; but this does not make their reports more effective. For example, the air pollution problem has been reviewed by countless committees and task forces. These have produced voluminous proceedings, usually written in a turgid prose that no one reads. The proceedings eventually find their way into incinerators, thus contributing to the increase in air pollution rather than to its control. Membership in a blue ribbon committee or a task force is at best a form of sublimation that provides a glamorous substitute for becoming involved in the practical affairs of the world.

Whatever the nature of their assignment, all committees and task forces inevitably emphasize the need for more knowledge. This self-evident advice provides the scientific community with still another kind of legitimate excuse for continuing the pursuit of its own self-selected intellectual interests. Fundamental research is always justified because it holds the promise of new information that might be relevant to the problem at hand, but it is also a form of escapism because it eliminates the necessity of facing the far more complex and more difficult matters of practical applications.

Concern for social problems too remote in time for immediate action is another form of escapism which provides safe intellectual titillation. Typical examples of this kind of pseudo-concern are the ethical problems that would emerge if scientists were capable of altering man's hereditary endowment. There is no evidence that such genetic manipulation will be done in the foreseeable future. Furthermore, even if it became practical, social limitations would prevent its being practiced on a scale large enough to affect the fate of human populations. But the ethical problems raised by the possibility of manipulating man's genetic nature are not made less intellectually titillating by the great likelihood that the issue will not come up in our lifetime. The discussion of these problems provides an excuse for neglecting really urgent issues that demand pointed studies and action *now* and therefore would disturb the comfort of academic life. It is known for example that the contamination of slum children by the lead in paint is responsible for a large amount of disease, particularly of mental retardation. Yet, very little is done to study this problem. Scientists find it much more entertaining to talk about such a far out possibility as the manipulation of the genetic code, than to concern themselves with the control of lead poisoning, a childhood disease which is a social crime of today.

Increasing numbers of scientists are explicitly acknowledging the need for immediate action on the present problems of mankind. Some of them are pessimistic concerning the possibility of a solution, as was Norbert Wiener when he wrote the following lines in 1962:

There is a real possibility that changes in our environment have exceeded our capacity to adapt. The real dangers at the present time—the danger of thermonuclear war, the computing-machine sort of danger, the population-explosion danger, the danger of the improvement of medicine (to the extent that we shall very soon have to face not letting people live as part of the policy of letting them live)—all of these dangers make one wonder whether we have not changed the environment beyond our capacity to adjust to it, and whether we may not be biologically on the way out.

This gloomy view of the future calls to mind H. G. Wells' words in "Mind at the End of its Tether": "Everything was driving anyhow to anywhere at a steadily increasing velocity. . . . The pattern of things to come faded away. . . . The end of everything we call life is close at hand and cannot be evaded."

Both H. G. Wells and Norbert Wiener were ill and approaching death when they wrote these gloomy statements. But many young and healthy scientists also express dismay about the present state of affairs in words that seem to imply "Stop the world, I want to get off." There is a hint of this mood in a speech (refer to page 64) that the astronomer Walter Orr Roberts delivered in 1967 at a meeting of the American Association for the Advancement of Science (AAAS), "We will likely know when the first intercontinental missile of World War III comes, should that happen, in a routine computerized check, on milli-second time scale, of the inventory of space debris; and the decision to retaliate, to enter total war, will probably be made on computer-based advice."

The need to supplement the orthodox scientific approach to human problems with one based on humanistic values was the keynote of another lecture delivered by Lynn White, Jr. before the AAAS in 1967, under the title "The Historical Roots of our Ecologic Crisis": "More science and more technology are not going to get us out of the present ecologic crises until we find a new religion, or rethink our old one. . . . The sense of the primitive Franciscans for the spiritual autonomy of all parts of nature may point a direction. I propose Francis as a patron saint for ecologists." White's advice that we turn to Francis of Assisi for scientific and technologic salvation amounts to a counsel of despair and a retreat from the world. Not surprisingly, the San Francisco hippies have made St. Francis their own patron saint. They even recognized White as one of their prophets after the publication of his lecture, going as far as reprinting it *in extenso* in their newspaper *The Oracle*!

White's recommendation is a form of escapism, because the passive relation to nature preached by the early Franciscans is now impossible to practice and furthermore never existed. Man has never been passive about nature; he has manipulated it and attempted to mold it according to his wishes ever since he acquired his biological identity during the late Paleolithic period. Among the founders of religious orders, St. Benedict would be a more suitable patron saint for modern scientists. His monastic rule dictated that monasteries be self sufficient. As a result Benedictine monks learned to manage nature on sound ecologic principles; they incorporated in Benedictine life a number of wise practices that related the external world to fundamental needs of the body and the mind.

There is not much chance in any case that we shall follow either St. Francis or St. Benedict, whatever admiration we may express for their teachings.

We dislike polluted and cluttered environments, but we like economic prosperity and gadgets even more. For this reason I doubt that we shall seriously undertake the social and technological reforms that would be essential for environmental control, until we are forced into action by major catastrophes.

Scientists and citizens who realize the gravity of the threats to human life posed by the undisciplined and asocial use of technology commonly acknowledge the need to put controls on unbridled economic expansion. This attitude takes several different forms. One, based chiefly on economic and esthetic considerations, was aptly expressed by the former Secretary of the Interior Stewart Udall in his article "Can America Outgrow the Growth Myth." Another form is the surprising statement by the illustrious physicist I. I. Rabi, Nobel Prize Laureate, that science does not play as large a role in technology as usually assumed (1965):

I'm not sure that science has been so terribly important for a lot of the basic technology we have today. Just thinking about it recently, I recalled the older arts and the great progress that was made without science: the wonderful metals, before the knowledge of chemistry or metallurgy was even developed; the wonderful fabrics that were made; building materials; sailing ships, really the most improbable and remarkable things in the world (much more so than steamships); techniques of mining, agriculture, and forestry; the arts of war without the help of RAND and the arts of statecraft without the help of Brookings; the preservation of foods; the utilization of water power, wind power, horsepower—I mean the tremendous advance that was made from the time of the ancients to the Middle Ages in the harnessing of horses and the domestication of animals; dyeing, weaving, pottery—all these things were developed to a very high degree without the help of science.

When we look at the history of industry, it seems to me that science had very little effect on industry, that is, on the older arts, until it developed to a very sophisticated point. (For a similar statement, refer to Seitz, page 80).

It is certainly true, as suggested by I. I. Rabi and F. Seitz, that science has not "been so terribly important for a lot of the basic technology we have today." And this appears to provide scientists with an excuse for emphasizing the theoretical issues that interest them most and not worrying about the technological consequences of science. On the other hand, it is also true, as both Seitz and Rabi acknowledge in the statements quoted on pages 80 and 128, that technology becomes more and more dependent on science as it becomes more sophisticated. Science and technology are now inseparable. For this reason, many scientists are now deeply concerned with the technological applications of their discoveries, as expressed by E. W. Engstrom: "The introduction of new technology without regard to *all* of the possible effects can amount to setting a time bomb that will explode in the face of society anywhere from a month to a generation in the future."

If taken literally, Engstrom's admonition would

encourage a new form of escapism, since it is certainly impossible to determine *all* the possible future effects of a new technology. A more guarded form of pessimism concerning the dangers posed by scientific technology has been expressed by Harvey Brooks, Dean of the Harvard Division of Engineering and Applied Physics, and one of the most influential leaders of scientific policy in the U.S.:

> Scientists and others who think about science and technology are becoming increasingly aware that decisions about science and technology must be made in the light of their possible second-order consequences—even when these cannot be anticipated—since the disadvantageous consequences of introducing a new technology can at times outweigh the primary expected benefits.
>
> The scientific community may thus become a much more *conservative* force in society than it has been, or may adopt an ambivalent attitude toward change. . . . [Italics mine] (Brooks, 1967)

The use of the word "conservative" in this statement gives the impression of a Freudian slip. It suggests that scientists are now more preoccupied with conserving the present, than they are with developing a creative approach to the future. If this attitude is widespread it implies a loss of nerve on the part of the scientific community. Yet, there is so much pioneering work to be done—beyond technologic solutions—to make science really serve human welfare.

I have discussed elsewhere some of the gaps in

our knowledge concerning the impact of scientific technology on human beings and on their environment. (Dubos 1965, 1968) Here, I shall select one particular aspect of modern life, the solid waste problem, that may appear of limited scientific interest yet will cause industrial civilization to choke to death if a systematic scientific effort is not rapidly organized for its solution.

Every ecologist knows that the ultimate objective in environmental control should be to manage society in such a manner that the products of its activities are recycled for re-use, instead of being wasted and contributing to environmental pollution.

Much scientific study will be required to develop the potentialities of recycling. A few examples of achievements will illustrate how recycling can make environmental control more effective by attacking waste and pollution at the source.

In some cities garbage is collected to make fertilizer through a biological process of composting. Dust from some grain elevators is made into pellets for cattle feed. In some plants fly ash from smokestacks is collected for making bricks and cement. Sulfur from a number of oil refineries and sulfur dioxide from factory chimneys is used as raw material for the manufacture of sulfuric acid. Admittedly in most cases the cost of the salvaging operation exceeds the value of the recovered material. But in the long run, and in terms of total science accounting, it will be more economical to recover wastes at the source than to clean the air, land, lakes, and waterways after the pollutants have been dispersed and have caused damage.

To be fully developed, the concept of recycling must include the manufactured goods after they have served the consumer. Real environmental control cannot exist until we develop technologies to recycle the materials from discarded equipment, such as automobiles, refrigerators, freezers, washing machines, vacuum cleaners, radios, and television sets. Recycling presents many extremely difficult scientific questions, but they should be regarded as the most urgent ones of scientific technology. If we are serious about improving the quality of the environment, we must establish centers where such problems are studied deeply by the most sophisticated scientific methods.

But we do not seem to be really serious about environmental control. We formulate the problem in orthodox economic terms, as if the staggering cost of present waste mismanagement was not much greater than the increase in the cost of controlling waste at its source. We accept waste and pollution whenever it makes technology financially more profitable, even if the products are useless or dangerous gadgets or substances.

In the course of their daily work, scientists naturally become increasingly engrossed with the problems of their own creation, and they tend to neglect the ones of more direct relevance to practical affairs. Although they plead for the expansion of the scientific enterprise and exploration of new fields of knowledge, often they behave as if they were interested in the production of more and more of the same kind of science. However mere growth in size, almost inevitably results in decay. To survive intellectually and socially, scientists must refocus their work on social objectives.

The greatest danger for the future of science is the present tendency to publicize its spectacular feats to gain public support. Extravagant and unwarranted claims are most commonly made concerning practical applications. But they are also made on theoretical subjects that have popular appeal—for example, the alleged imminence of producing life synthetically in the test tube and communicating with unknown creatures assumed to populate distant celestial bodies.

Unwarranted statements not only are intellectually dishonest, they also are dangerous because they encourage the belief that science will soon explain all mysteries of the cosmos and solve all practical problems by inventing new technologies and countertechnologies. Scientists will find ways to clean the air over urban areas and purify the water of streams and lakes; so why enforce measures against pollution of the environment? Scientists are inventing methods for increasing food production, discovering new materials to replace the ones that are becoming scarce, and developing settlements on the moon or the floor of the oceans; so why be concerned about overpopulation and the exhaustion of natural resources?

Since scientific evidence can be evaluated only by specialists, every person tends to select, among the pronouncements made by the various scientists, those interpretations which fit his own prejudices: that nuclear explosions threaten man's genetic inheritance or that they do not increase significantly the natural occurrence of radiation damage; that the statistical chance for the emergence of life from matter is practically nil or that life has emerged repeatedly and continues to emerge in many parts of the cosmos; and that the universe began at a given time with a big bang as by an act of divine creation or that it is without beginning or end and is continuously being created. Each person can bolster his own prejudiced opinion on these and other unsettled matters by quoting a particular scientist who has stated as an established fact what is at best an interesting hypothesis. Unwarranted statements mislead the public and encourage it to avoid the effort and discipline of critical thought.

Scientists are betraying the spirit and ideals of their profession when they accept the mores of the marketplace. To deserve and regain public confidence they must rededicate themselves to the kind of intellectual integrity that makes a clear differentiation between what is unequivocally known and what is merely hypothetical. Instead of being satisfied with expanding their activities along fashionable lines, they must try to discover which fields of study are most likely to yield insights and provide guidance for social action.

Modern society will continue to believe that scientists should be "on tap but not on top" as long as the scientific establishment does not seriously try to formulate and foster the kinds of socially oriented knowledge that will help man live intelligently in the technological world.

Future-Oriented Institutions

In the late 1930's a progressive and prosperous industrial firm in the United States created a research institute for the study of scientific problems relevant to its commercial interests. To glamorize the dedication of the new institute the firm invited illustrious scholars to discuss the organization of scientific research. The major address was given by the late Abraham Flexner, now remembered for his historic report on the reform of medical education, the great influence he exerted in shaping the policies of the Rockefeller Foundation, and the creation of the Institute for Advanced Studies in Princeton.

The title of Flexner's address was "On the Usefulness of Useless Knowledge"; his views of scientific research were based on the following assumptions:

(a) Scientific knowledge develops as an autonomous process according to a logic of its own, essentially uninfluenced by social events or needs.

(b) Unexpected discoveries often occur by accident through the operation of serendipity. The word "serendipity" was created by Horace Walpole, and popularized by the American physiologist W. B. Cannon to denote the discovery of things or facts that were not the object of the search (Remer, 1965). In his book *The Way of an Investigator*, Cannon defended the view that serendipity was favored by completely free, unplanned research.

(c) Granted these assumptions, it follows that scientists should be unconcerned with factors external to science in the selection of their problems and should be given as much freedom as possible in the prosecution of their studies. Scientific programs should not be directed or evaluated by outsiders, because scientists work most effectively when they follow their personal logic and are free to seize on any accidental observation made by serendipity.

(d) The most esoteric and apparently useless kinds of knowledge are very likely to yield unexpected practical applications. Furthermore, these applications commonly find their place in human problems and activities that are unrelated to the fields of science in which the theoretical studies were made. Laissez-faire, not planning, should be the philosophy of organization for scientific research.

The assumptions that underlie Flexner's essay on the "Usefulness of Useless Knowledge" are naturally congenial to most scientists. In fact, they have been widely accepted not only by scientists but also by enlightened laymen. For this reason, they have exerted a dominating influence on the administration of scientific research and account for the general reluctance to organize it and to give it direction.

Names such as Roentgen, Becquerel, Curie, Einstein, and Fleming have been mentioned ad nauseum to illustrate that complete laissez-faire is the most profitable policy in the administration of scientific research. The immense contributions of science during the Second World War have been cited as further evidence that intellectual freedom is essential to scientific creativity. Indeed it is true that most of the knowledge applied to the development of weaponry and medical procedures during the war had its origin in basic research carried out in uncommitted academic laboratories before the war.

Despite this evidence Flexner's philosophy of scientific research seems to be inadequate on two counts. It is based on a faulty reading and interpretation of the history of science. It fails to take into account social preoccupations and implies a passive attitude toward the needs of the future. I shall build my argument chiefly from examples pertaining to the biological sciences because this is my area of professional specialization. However, similar examples could be taken from mathematics, physics, or chemistry to illustrate that, although the Flexner philosophy has a limited application to the *tactics* of scientific research, it is invalid when applied to the overall *strategy* of science.

The development of any particular field of science is profoundly influenced of course by forces inherent in the scientific enterprise itself; these will not be discussed here. Conversely, social forces unrelated to the logic of science play a large role, probably the dominant one, in determining which fields of science are emphasized at a given time and which ones are neglected.

One period of medical history illustrates how

social forces give direction to the interests of scientists and therefore to their research activities. Medicine first became scientific through the development of anatomic, physiologic, and biochemical sciences during the first half of the nineteenth century. At that time these kinds of medical knowledge had the greatest appeal for physicians, scientists, and laymen. When Claude Bernard published his immensely influential book *Introduction to the Study of Experimental Medicine* in 1857, everyone was convinced that physiology would revolutionize medical practice and therefore would be the dominating force in medical sciences. Yet, the first great medical research institutes were not organized around physiology; they were devoted instead to the bacteriologic and immunologic sciences, because the most common and most devastating medical problems of the nineteteth century were caused, or at least aggravated, by microbial infections.

Neither medical schools nor universities were geared for the study of microbial diseases during the nineteenth century. Special institutions therefore were created for the prosecution of sciences focused on these diseases. In France, the Pasteur Institute was set up in 1888 as a private corporation with an endowment obtained largely from voluntary subscription. In Germany two separate institutes were established by the government, one in Berlin for Robert Koch and the other in Frankfurt for Paul Ehrlich. In England and Japan, private philanthropists financed institutes to continue the work begun by Joseph Lister and Kitasato. Whatever the country, the scientists in whose name medical research institutes were created had achieved fame in some problem related to the understanding and control of microbial diseases. The first institutes devoted to medical research thus emerged independently of the academic establishment; they were established and nurtured by private and political bodies to deal with an urgent social need.

The institutes of medical microbiology (founded three generations ago) have continued to function in our times, but their relative importance on the medical scene has progressively decreased because microbial diseases are now less frightening than they used to be. In contrast, research laboratories devoted to the study of biochemical, physiologic, and genetic phenomena have been created all over the world because metabolic and degenerative disorders have become more prominent.

Early in the present century, medical scientists as well as a few enlightened physicians and laymen realized that progress in medicine would benefit from greater knowledge of theoretical biology, chemistry, and physics. This recognition pointed to the need for enlarged concepts of biomedical research and for new kinds of scientific specialists. From this awareness the Rockefeller Institute for Medical Research emerged in 1903. The purpose of its founders was to create facilities and an intellectual atmosphere in which investigators could dedicate themselves to the development of physical, chemical, and biological sciences relevant to knowledge of the human body and to the control of its diseases. This broad formula of medical research has been incorporated in the structure of modern medical schools all over the world. In fact, its acceptance is so universal that around 1955 the trustees of the Rockefeller Institute considered that its initial mission had been fulfilled. Therefore, they felt that the Institute could more profitably devote itself to other scientific problems of social interest, such as the development of a stronger basis for the study of behavior and the improvement of postgraduate education. The Rockefeller Institute for Medical Research thus rapidly evolved into The Rockefeller University.

The change in emphasis from microbiologic to chemical, physiologic, genetic, and then to behavioral sciences has a logic which is not inherent in science itself, but is derived from social concerns. The history of science provides many other examples of such shifts in scientific emphasis. These often occur rather suddenly because they are brought about, not by the logical evolution of knowledge but by forces originating in the social environment.

Some of the forces that give its direction to the scientific enterprise are political in origin—as illustrated by the proportion of public funds allocated to the various fields of science since 1940 and especially post-Sputnik.

Most human beings are more interested in their biologic and cultural origins than in the exploration of subatomic physics or outer space; this is shown by the comparative frequency of topics of conversation among the general public (including scientists), dealing with man's origins; the sale of books written for laymen also proves this interest. Moreover, there is no evidence that knowledge of subatomic physics or of outer space is theoretically more important or useful than the knowledge of man's nature or cultures. The decision of governments to support the former and not the latter is based almost entirely on considerations of power politics and international prestige, rather than on intellectual or humanitarian principles. Neither the logic of science nor the public taste had any part in this decision.

Yet other fields of science have been developed as a result of public pressures, despite the reluctance and even the active opposition of the academic community. A century ago the agricultural experiment stations and the various organizations grouped around the land-grant colleges (Morrill Act, 1865) were established in each state to provide a scientific basis for practical agriculture. This new enterprise which originated from the efforts of farming interests acting through Congress was long regarded with contempt by orthodox academics. Yet the agricultural experiment stations and the so–called cow colleges rapidly evolved into great centers of learning, which have advanced not only the practice of agriculture but also many other fields of theoretical knowledge.

The shifts in directions of scientific research brought about by social forces would be of little importance if it were true, as implied in Flexner's essay and affirmed by the devotees of serendipity, that great advances in science and in its applications are commonly the unexpected results of accidental findings. However, the statement that discoveries happen more or less by chance is at best a half-truth. It has its origin in the least intellectual aspects of the Baconian tradition and is, in fact, anti-intellectual.

The belief that scientific planning is useless or even dangerous because *any kind* of research eventually leads to practical applications is also false. I shall limit my remarks on this topic to an anecdote taken from recent medical history.

The introduction of penicillin into medical practice is repeatedly quoted as proof that great practical values often come from uncommited research and accidental discoveries. Fleming's isolation of his famous mold seems indeed to constitute the perfect demonstration of the usefulness of unplanned, apparently useless knowledge. The details of the penicillin history, however, point to a different interpretation. (Maurois, 1959)

For many years before he discovered penicillin, Fleming had been working with antibacterial substances and had studied their role in infectious processes. This interest had grown from his association with Almsworth Wright in whose laboratory he had begun his scientific career. Fleming was intellectually and even emotionally conditioned by his early scientific experiences to welcome the mold when it fell accidentally on his workbench, contaminating a staphylococcus culture. One of his first experiments after he recognized the antibacterial activity of the crude filtrate prepared from the mold was to test its toxicity for various blood cells. His tests for antibacterial activity and toxicity of the new product were direct extensions of the tests he had carried out many time before with other antiseptics.

Fleming would not have noted the presence of the mold, or recognized its potential usefulness, if he had been working on intermediate metabolism, the ultramicroscopic structure of collagen, the transmission of nerve impulses, or any scientific problem other than antibacterial activity; he discovered penicillin because he was interested in microbial diseases, particularly in substances that inhibit microbes.

Fleming lacked the knowledge of chemistry needed to convert his crude mold filtrate into a usable product, and for this reason his discovery was almost forgotten for ten years. Nevertheless as a result of their concern with the treatment of combat wounds H. Florey and E. B. Chain revived the penicillin phenomenon in 1940 and worked to make it practically available to clinical medicine. The contribution of these two investigators had its origin in an awareness of the acute need for new antibacterial drugs—an awareness which encouraged them to develop practical techniques for the production of penicillin from Fleming's crude filtrate.

Penicillin was indeed discovered by bench serendipity, but the initial finding was made, and the scientific knowledge required for its use was developed, because the scientists concerned had well-defined goals in mind. "Chance," Pasteur wrote, "favors only the prepared mind." The mind must be prepared not only by scientific training and technical proficiency but also by the awareness of social needs.

The research institute dedicated by Flexner in 1939 was founded by a pharmaceutical company. It has been highly productive in the discovery and utilization of drugs and hormones--the products sold by the firm—but it has not contributed much to other areas of medicine. It has not provided the right environment for other kinds of discoveries nor has it tried. This was not its mission.

Scientific research is never entirely free and uncommitted. To be effective a research group has to specialize in certain fields of science and neglect other fields. Areas of emphasis are determined by professional interests, availability of funds, public pressures, and, of course, economic necessities in the case of industrial enterprises.

University scientists themselves are much more susceptible to para–scientific influences than they are willing to admit. Pure intellectual curiosity and concern for large theoretical issues motivate some of them, but the selection of research problems predominantly has many other determinants, among them the character of the institution in which the scientist works, the prevalent scientific fashions, and the emergence of new social needs. For example, during the past two decades the majority of academic biologists have acted on the current assumption that the study of organic macromolecules constitutes the most important and most urgent task of biology. This attitude, however, may soon become outdated, if one judges from the following statement by one of the most eminent and respected leaders of American biomedicine:

> Consider the study of the biological, behavioral, and intellectual development of the child as an end in itself, with a view to coping ultimately with mental retardation and assuring healthy intellectual growth. Here, we are beginning to explore environmental factors with a wholly new social orientation. Our national program in support of science is entering a phase in which the decision whether to support a given field depends less upon technical consideration than upon social need (Shannon, 1966).

In science as in other human endeavors, no man is an island.

The preceding remarks are not meant to imply that there is no place in scientific research for uncommitted curiosity and that only mission-oriented research is useful. Knowledge is useful even if it only enlarges man's understanding of himself and his environment, thus helping him to make his life more intelligible and enjoyable. Consciously or subconsciously, however, scientists follow directions that are determined in part by social influences. Therefore we must develop a more sophisticated knowledge of human needs and desires if we want science to fulfill its social role. Since even the most prosperous country cannot possibly have enough resources, especially talented and trained personnel, to study all the scientific problems that could be profitably investigated, the recognition of social needs and desires implies the establishment of priorities. These are largely determined on the basis of para-scientific criteria which usually involve subjective anticipations of the future. The social goals of science should never be set irrevocably, because the future will certainly be very different from what we can imagine. As will be pointed out later, planning not only involves predicting the future but also structuring social institutions in such a manner that goals can be rapidly reformulated to avoid dangers that had not been predicted and to realize unforeseen opportunities.

Most would agree that science and technology are responsible for some of our worst nightmares and have made our societies as complex as to be almost unmanageable. A few scholars agree with the hippies that the social problems of the modern world are created by science and cannot be solved by more science (White, 1969). It is beguiling to hope that salvation can come simply from a change in social philosophy, for example, to reject bourgeois conventions and return to the love of nature practiced by the early Franciscans. Most persons, however, take a more prosaic view. While acknowledging the limitations of the technological world they enjoy it nevertheless and hope that science will provide ways to manage it more intelligently. Our ancestors mobilized science a century ago to master nature and to create wealth through technology. We are now asking science to save us from technology. The report of the Daddario Subcommittee in the U.S. House of Representatives (1966) expresses the apprehension and hopes of the public: "The subcommittee believes that we cannot blindly adapt technology to our needs with the traditional assumption that there will be ample time to iron out any bugs on a leisurely shakedown cruise. A bigger effort must be made not only to foresee the bugs but to forestall their development in the first place."

The need to focus scientific effort on the social problems of the modern world is affirmed in the reports of countless committees and task forces. These efforts, however, will remain daydreams until academic institutions transform themselves or new kinds of research institutes are developed.

Universities are the great reservoirs of classical knowledge and still constitute the most important source of new knowledge. But the kind of scientific research they are best equipped to carry out is not likely to provide the specific technical information needed for the improvement of modern societies.

The academic structure was organized for undergraduate education and later modified to provide discipline-oriented training for predoctoral students. This educational task remains its primary responsibility, and academic institutions tend therefore to perpetuate themselves, through self selection, in the image of their forebearers. In theory it might be possible to keep the academic establishment sufficiently loose and unstructured so that faculty members could become involved in practical problems of social concern whenever they feel so inclined. In practice, however, universities tend to maintain the status quo, because academic scientists investigate those aspects of the social problems that fit their own professional interests.

The academic investigator insists—as he should—on the right to pursue his own interests regardless of their social relevance. He tends to be interested chiefly in the inner logic of his scientific discipline and is rarely "mission-oriented." Alvin M. Weinberg has recently pointed out that, "As the disciplines making up the university become more complex and elaborate in response to their own internal logic, the discrepancy between the university and the society grows. The university becomes more remote; its connection with society weakens; ultimately it could become irrelevant."

Academic research naturally yields knowledge which often has practical value, but many scientific problems of great social importance remain outside its scope. Even medical schools tend to neglect some of the largest health issues of our time. Like university professors, investigators in academic medicine want to follow the inner logic of their scientific interests, and this often takes them far away from many health problems that do not appear to them of deep theoretical interest yet are very important for the community. As a result of this academic attitude much of medical research is almost indistinguishable from the kind of biologic investigations carried out in nonmedical environments.

Furthermore, the scientific study of most social problems requires a complex, integrated approach not readily achieved within the present academic structure. All successful examples of scientific integration in the university are provided by research units focused on special areas of basic science such as geophysics, high energy physics, or molecular biology. These units, which are made up of scientists working in different fields, appear somewhat outside the classical university system, because they are focused and coherent in their approach to research projects. Nevertheless, their outlook is so specialized and remote from practice that they cannot deal effectively with the immediate problems of man in the modern world. They are concerned primarily with science for the sake of science rather than with science in human affairs.

Professional associations have long realized that it is difficult to carry out the scientific study of their problems within the classical academic system. For this reason they have established their own research programs outside the universities. The agricultural experiment stations, the School of Aeronautics at the California Institute of Technology, and the institutes devoted to communication theory in the laboratories of the Bell Telephone Company or of the IBM Corporation are but a few of the numerous examples of this trend.

In medicine, also, it has become apparent that the research attitude and methods suited to the study of general biologic phenomenon are not sufficient to deal with many of the problems that preoccupy the modern world. When these problems become pressing, independent institutes are created to deal with them, as was done in the past for the control of microbial diseases. For example, academic physiologists may choose to focus their studies on the phenomena of conduction in isolated nerve fibers, but large research units are being created outside medical schools to investigate the special physiologic problems of aviation and space medicine. Similarly, departments of biochemistry in medical schools may emphasize the intricacies of intermediate metabolism, but special institutes for the study of malnutrition and of its multifarious effects on human life are emerging in underprivileged countries.

The indirect and delayed toxicity of drugs has become a matter of much concern among physicians, but few academic institutions encourage research in this field. The distribution of medical care constitutes another area in which there is great demand for fundamental investigations of man's biologic and social needs. The lasting effects of physiologic and psychic experiences during the early phases of development also present problems that are much neglected—even though it is now certain that early influences condition practically all physiologic and mental characteristics of the adult; early influences also play a dominant role in all pathologic states ranging from obesity to drug addiction. The remote effects of environmental pollution similarly pose problems of immense magnitude for the study of which universities and medical schools are poorly equipped, both physically and intellectually.

It might be argued that the ease of communication will make it unnecessary to create new institutes for the study of special problems. The Neurosciences Research Program, which involves the collaboration of many investigators who work in different parts of the world and meet at frequent intervals, provides a pattern that will certainly be imitated in other fields. This type of pan-institutional organization, however, is not likely to be suited to the study of complex social problems. Increasingly, the scientific knowledge needed for a more reasonable conduct of human affairs will have to be acquired in mission-oriented research institutes.

The development of mission-oriented science will certainly be influenced by the history of the agricultural experiment stations and of the land-grant colleges associated with them in the United States. The very emergence of these institutions shows that public pressure (in this case from the farming interests) can hasten or even generate the development and teaching of neglected fields of science. The success of the experiment stations in improving the practice of agriculture was due to the fact that they facilitated a rapid feedback between practical problems faced by the farmers and the knowledge of laboratory scientists. The feedback was accelerated by extension services in which practitioners could formulate questions and take the theoretical answers to the field. When such answers were subjected to the test of practical conditions, new theoretical problems were recognized which were taken back to the laboratory for further investigation, and this enriched knowledge.

The interplay between theoretical knowledge and practical application is not a new phenomenon. In fact, it may well have constituted the most important factor in the history of science. However, this interplay is acquiring special importance at the present time because of its relevance to the development of scientific policies and social planning.

Organizing scientific research on the basis of social and economic criteria is commonplace in large industrial firms, but in governmental institutions it is only beginning. William D. Carey, formerly Assistant Director of the Bureau of the Budget and now with Arthur D. Little, Inc., has attempted to make a comparative evaluation of competing research programs on the basis of their relative contributions to the different types of national goals. A tentative example of such evaluation is outlined in the table on page 152.

Of course it will be difficult to achieve a consensus on the relative importance of various goals included in the matrix. As emphasized by Carey (private communication): "Anybody's formulation of a social merit matrix is bound to be based by the way he judges society's perceptions . . . the value categories could be representative of *apparent* social preferences without penetrating the much more volatile *suppressed* preferences below the surface." The supersonic transport plane (SST) provides an important test case, because it illustrates how conflicting social criteria complicate the evaluation of scientific technology.

According to an advertisement paid for by a spokesman for aviation, "Supersonic supremacy is the absolute condition of America's future security. . . . It must grow with major advances." This statement does not promise that supersonic planes will confer any new blessings upon us; it only promises that they will enhance national prestige—whatever this may

Research Programs and National Goals

Value Category	Weight	Desalination	Population Control	Weather Modification	Oceanography	Lunar Exploration
ECONOMIC						
Health and Welfare	3	x	x	x		
Technological gain, business expansion, full employment	10			x	x	x
Conservation of resources	10	x	x	x	x	
Return on investment (cost-benefit)	2	x		x	x	
CULTURAL						
Exploration	5				x	x
Understand environment	5			x	x	x
Enrich education	10					
Improve human relations	5		x			
POLITICAL						
National prestige	2	x				x
International understanding	5				x	
Problem-solving in underdeveloped countries	3	x	x	x		
Cold war advantage	15			x	x	x
Relative Program Values		20	21	48	52	37

mean. Other groups, however, are defending the view that the price to be paid for any advantages gained from supersonic speed will be far too high. They emphasize that the hours of flight saved by the few thousand officials, business tycoons, or the idle rich who will fly on the SST will not compensate for the disturbances caused to the millions of people exposed to the sonic boom on the ground. Many other conditions of supersonic flight have scientific determinants which must be evaluated in the light of social values.

All political decisions involve similar scientific dilemmas. For reasons of national security and perhaps even more, international leverage, the United States considers it essential to maintain nuclear superiority over the rest of the world. However, while this is being achieved through tremendous expenditures of money and scientific talent, among the countries of Western civilization the United States has one of the highest rates of infant mortality and lowest life expectancy for males over forty-five years of age. Little is being done, financially and scientifically, to lower death rates in these age groups because health standards are judged of minor importance when international prestige is at stake.

Scientific research is never entirely value-free because science needs the support of society and affects most social decisions. The large amount of talent, time, and funds devoted to any one program necessarily limits the support that can be given to another program.

The greatest obstacles to the organization of

mission-oriented scientific research programs may not be administrative but will come from the need to develop new attitudes toward scientific philosophy. Throughout most of the world the typical scientist is trained to think of science not as a means but as an end; his professional ideal is science for science's sake. If social problems are considered from the purely scientific point of view, scientists are—*ipso facto*—the best judges of their relative importance and optimal solution. However, this is not the case when the same problems and their solutions are considered from the point of view of social relevance.

Mission-oriented research has objectives that are societal rather than scientific. Social rather than scientific criteria determine what kinds of problems are important and what technical solutions are most appropriate.

Scientists will not readily reconcile themselves to this change of emphasis. Their unhappiness will not come from a lowering of scientific standards or a scarcity of interesting problems. In fact, applied science is in many cases more demanding intellectually than are the so-called purely academic sciences. The investigator in applied science must accept the complexities of the natural world instead of selecting problems on the basis of their convenience for experimental analysis, suitability for rewarding speculation, or opportunistic appeal.

Scientists have learned from long experience, however, that they are happiest and most productive when they have the freedom to select their areas o activity, to work according to their temperament, and to follow whatever interesting phenomena turn up in the course of their studies.

The shape of things to come may be seen in a recent analysis of the present and future role of National Laboratories in the United States. A. M. Weinberg, who has guided the research course of the Oak Ridge National Laboratory for more than twenty years, points out that mission-oriented scientific establishments such as those of the Atomic Energy Commission age rapidly because their missions are completed. Their laboratories need new assignments to retain their vitality. The word assignment, however, implies an attitude toward research different from that held by the majority of academic scientists.

One may anticipate that the scientific structure will evolve even further than suggested by Weinberg, and that much mission-oriented science increasingly will be carried out by task forces made up of assorted scientists possessing diverse skills and organized around unsolved problems. Temporary project groups rather than permanent stratified units would characterize this type of mission-oriented science. The necessity to function in temporary work systems and to cope with rapid changes would certainly create strains that present-day scientists, working either alone or in small, stable groups, have much less occasion to experience.

In addition to the emotional disturbances that mission-oriented research is bound to cause scientists, there are other kinds of potentially dangerous changes that are likely to occur in the scientific process itself.

The formulation of objectives determined outside the research groups on the basis of non-scientific criteria may rob science of its own self-generating intellectual creativity and excitement. In the past, this very excitement did much to generate objectives that were both worthwhile and realizable.

All discussions of science planning are inevitably naive, because we know so little concerning the social workings of the present scientific structure and the manner in which scientists will respond to the various forms of organization and control that are now being contemplated. It can be anticipated, however, that working scientists will function most effectively in mission-oriented institutions if they have a chance to become intimately involved in the formulation of the social objectives to which their work will contribute. The dynamic, on-going character of the scientific enterprise may well give scientists the opportunity to prevent long-range planning from slowing down social evolution.

Whatever type of teleological thinking and long-range planning emerges, social goals should never be set irrevocably. Plans have to be altered when unexpected difficulties obstruct or depreciate the original objectives. More importantly, perhaps, goals should be altered when new facts and new outlooks point to more desirable goals.

Social institutions fulfill two different but complementary roles. One is to promote stability of purpose amid the disruptive pressure of daily events; the other is to facilitate and accelerate adaptive responses to changes in the environment and the characteristics of the social group. Universities and research institutes have been remarkably successful in assuring the intellectual stability required for the long-term effort which has produced our scientific technological civilization. In contrast, they have dealt much less effectively with the new problems created by this form of civilization, and they have done little to explore the consequences of biotechnologic changes for man's future.

Admittedly, a few university programs are attempting to devise tactics for anticipating the consequences of the exploitation of the earth's resources so that we do not have to face, astonished and unprepared, a succession of environmental crises. This kind of prediction, however, is not sufficient to deal with problems of a future-oriented, rapidly moving society. We must also recognize the continuously evolving potentialities that are created by surroundings and events.

Universities have been established to transmit classical knowledge and to acquire new but timeless and universal knowledge. Research institutes as presently constituted deal with the problems of the present and of the predictable future. Now that the technologic environment is changing at an increasingly rapid rate, we must develop scientific institutions to study how man can best make use of the possibilities among which he can choose. Only then can he determine his long-range future, not as a passive witness but as a willful creator.

5
Scientific Civilization

The Technological Environment

In general usage the word "civilization" refers to values and practices that could apply to all people regardless of race, origin, or religion. Certain ethical concepts, much of scientific knowledge, and the ways of life derived from technology at its best, can be shared and enjoyed by mankind and thus can contribute to the development of a universal civilization.

Although civilized life can be traced far back in time and even in preliterate history, the word "civilization" is of recent origin. It seems to have appeared in print for the first time in 1757 when the Marquis de Mirabeau used it in his essay "L'amy des hommes ou traité de la population." Mirabeau entitled another essay that remained unpublished "L'amy des femmes ou traité de la civilisation." As the title of this essay suggests, Mirabeau believed that the development of civilized life was chiefly due to the influence of women (Nef, 1961).

Throughout the eighteenth century the word "civilization" had a far more restricted meaning than we give it now. It referred to gentle ways of life, humane laws, limitations on war, a high level of purpose and conduct, in brief, all the qualities considered to be the highest expressions of mankind. Nevertheless, as late as 1772, Samuel Johnson refused to enter the word in his dictionary because he felt that it did not convey any concept that was not covered just as well by the older and better defined word "civility."

Samuel Johnson may have been correct in rejecting the word "civilization" if one considers that it is now used to denote several types of human endeavor that are very different and indeed unrelated. For example, the classical civilizations of rational Greece or of artistic Florence had little in common with the industrial civilizations of the satanic mills or automated life. On the other hand, the very diversity of civilizations, testifies to the breadth of human potentialities. For at least 100,000 years man has used his beliefs, tastes, and hopes to fashion from natural forces the multiple components of civilized life.

The conscious processes that have enabled human life to become different from animal life were already at work during the Old Stone Age, long before the neolithic revolution. In fact, there cannot be any *human* life without civilization. Scientific civilization, however, involves much more than the various technologies used to manipulate the physical world, because intervention is conditioned by the cultural traditions and social practices that shape man's mentality and govern his behavior. Civilization is scientific to the extent that it is based on knowledge not only of things but also of man's nature. In our own time, however, science is shaping civilization much less by its effect on man himself than by the power it gives him to create the environment in which he develops. Some of the worst aspects of the modern world have their origin in the fact that we have not developed a science of civilization to complement technological sciences.

The present state of man and his environment makes it obvious that the use of scientific knowledge for the manipulation of nature does not necessarily produce happy results, but this awareness is relatively recent. Most philosophers and scientists of the Enlightenment and most technologists of the nineteenth century were euphoric about man's future and took for granted that health and happiness would *inevitably* be enhanced by scientific knowledge. They believed that an era of better human life—that is, of greater civilization in the nineteenth century sense of the word—was at hand because scientific advances would increasingly enable man to manage his own nature as well as external nature and that he would thereby become master of his destiny. If our ancestors had some doubts about the future, they were doubts not about technological developments but about human relations. At the end of the letter, mentioned earlier, to his friend J. B. Priestley, Benjamin Franklin expressed these doubts in the following words: "The rapid progress true science now makes occasions my regretting sometimes that I was born so soon. . . . O that moral science were in as fair a way of improvement than men would cease to be wolves to one another, and that human beings would at length learn what they now improperly call humanity."

Science and technology have now achieved much more than could have been anticipated from the most optimistic imaginings of our ancestors. Yet, it is obvious that something has gone wrong during the past century, not only with human relationships as Franklin had feared, but also with technological developments themselves. Increased control of nature has not provided safety and peace of mind; economic prosperity has not made affluent people healthier or happier; technological innovations create problems of their own that endlessly require the development of new countertechnologies.

As presently managed, industrial technology is rapidly destroying the elements of nature that have made possible the biological values and social amenities essential for humanness. The phrase "technological environment" denotes with increasing frequency pollution and congestion, exposure to deleterious stimuli, and the thousand devils of the ecologic crisis. Furthermore experience has shown, that practically all sudden environmental changes, so common in technological societies, are likely to disturb biologic and mental health.

Of course there is more to human ecology than the distressing aspects of man's interplay with his environment. The body and the mind are molded by the adaptive responses that man makes to the stimuli that impinge on him from the time of conception to the time of death. Modern man is being shaped by technological forces and thereby is becoming adapted to them. In theory, therefore, a balanced discussion of the environment should emphasize its formative effects as much as its pathological effects. But the environmental crisis must receive priority, because the earth will not long remain a place fit for human existence unless present trends are rapidly reversed.

It is commonly assumed that urbanization is responsible for our most serious environmental problems. In reality, there is no evidence that city life, per se, is less favorable to biological and mental well being than rural life. The pathological conditions which plague city dwellers are not inherent in urban life but result from its mismanagement; furthermore, most of these conditions are now just as prevalent and severe in rural areas as in urban agglomerations. Problems of life expectancy and disease are much the same in New York, London, Paris, Moscow, as among farmers, lumberjacks, or fishermen in any part of the world where their activities have become subjected to technology. In the city or the country, it is the technological environment which is chiefly responsible for the conditions that threaten biological and mental health as well as the quality of human interactions.

One can anticipate that industrial, economic, and medical difficulties will soon result from the rapid depletion of some of the natural resources on which technological civilization depends; even water will soon become scarce in the temperate zones. A more immediate threat, however, is that practically all environments are becoming grossly polluted. Pollution and depletion once were localized, but now they affect the whole earth. Deleterious and often destructive effects can be observed simultaneously everywhere over the earth's crust, in the atmosphere and in the various bodies of water; they affect practically all living organisms and naturally man himself. (Commoner, 1966; see also the new journal *Environment*)

Smogs and invisible fumes produced in industrial and urban areas are hovering over the countryside and beginning to spread over the ocean masses. Air pollutants formed in Texas have been detected more than 1,000 miles away in Ohio; pollutants originating in Germany and Great Britain reach Scandinavia under certain meterological conditions. In Sweden, air pollution has caused an increase in acidity of inland waters and thus injured fresh water fisheries. The urban atmospheres will not benefit much longer from the cleansing effect of the wind because the wind itself is now contaminated.

Like air pollution, water pollution is all but universal. Sewage, run-offs from heavily fertilized fields, chemical effluents, and pesticides are spoiling rivers, lakes, and beaches. Even the most carefully protected water supplies are slowly but surely being contaminated by countless chemical substances of industrial and domestic origin, many of which are likely to produce cumulative toxic effects. Recent observations in Sweden and Japan have revealed that mercury from various technological processes is accumulating in fish, often killing them or rendering them unfit for human consumption. Perhaps of most long-range importance is the fact that contaminants are beginning to concentrate in the plankton and may thereby upset the whole economy of the oceans and even the world supply of atmospheric oxygen.

For the sake of brevity, two more examples will be mentioned to illustrate that most pollutants are progressively achieving world-wide distribution. Insecticides such as DDT accumulate in insects, earth worms, plants, and ocean plankton, killing the fish and birds that feed on contaminated materials. The dramatic declines in several bird species such as eagles, falcons, and seagulls may be due to the accumulation of insecticides, which has decreased the thickness of their egg shells to the point of breakage. As to the potential effects of insecticides on human beings, it is sufficient to mention that the concentration of DDT now exceeds twelve parts per million in the fatty tissue of United States' residents. This may soon have tragic consequences, because DDT enhances the production of certain liver enzymes that destroy steroid hormones.

Radioactive products, not only from nuclear explosions, but also from nuclear power plants, are spreading throughout air, water, soil, and living things, and are eventually accumulating in certain food chains. For example, radioactive products from the fallout of nuclear explosions in isolated islands of the South Pacific are carried by winds to the arctic, where they are absorbed by lichens; from the lichens they reach reindeer that feed on these plants and eventually the Lapps and Eskimos who eat reindeer meat.

All aspects of modern life contribute to pollution. Bottles, tin cans, plastic containers, discarded machines, and other solid wastes and nondegradable forms of garbage litter the landscape wherever human beings have settled. New substances of untested toxicity are constantly being introduced into daily life. Excessive sensory stimuli and especially the mind-bewildering noise are difficult to avoid.

The complexities resulting from technological advances and crowding inevitably engender social regimentation, loss of privacy, and other interferences with individual freedom. Subliminal suggestions, hidden cameras and eavesdropping mechanisms for secret surveillance, and computers for recording and disseminating personal data are gradually becoming part of government and business practices. The computer networks encode more and more information about our private lives, and we are progressively accepting as a matter of course the sophisticated spying made possible by electronic circuitry and microminiaturization. The worst aspect of the takeover of life by technology may be that secret surveillance will eventually emasculate our civil liberties and cause them to be replaced by formalistic rituals.

It is possible of course to devise protective techniques against the environmental threats of modern life, but man cannot protect himself against everything all the time. We might find comfort in the fact that the human species, throughout its long biological history, has been able to adapt to many different kinds of environment and to survive under very difficult conditions. However, biological adaptive processes require long periods of time, whereas profound environmental changes now occur almost simultaneously, within a few years, and far too rapidly to allow for biological adaptation.

That modern man is now moving into non-terrestrial environments might be interpreted as evidence that he has escaped from the bondage of his evolutionary past and is becoming independent of his ancient biological attributes. But this would be an erroneous interpretation. The human body and brain have not changed significantly during the past 100,000 years, and there is no ground for the belief that they will change appreciably in the foreseeable future. The biological needs, capabilities, and limitations of modern man are essentially the same as those of the Paleolithic hunter and Neolithic farmer. What we now call civilization provides man with techniques that greatly enlarge his power and the scope of his activities, but it does not change his fundamental nature.

Wherever he goes and whatever he does, in tropical deserts or arctic wastes, in outer space or ocean depths, man must maintain around himself a microenvironment similar to the one within which he evolved. He may establish stations in outerspace or on the bottom of oceans, but few persons will live in them. He will have to function within enclosures that almost duplicate a Mediterranean atmosphere, as if he had to remain linked to the surface of Earth by an umbilical cord. He may engage in casual flirtations with planets and stars, but he is wedded to the crust of Earth, his sole source of sustenance.

The strict dependence of the human organism on the narrow range of terrestrial conditions imposes inescapable constraints on civilized life. In practice, social and technological innovations are viable and successful only to the extent that they are compatible with the unchangeable aspects of man's nature. Man will retain his biological and mental health only if he learns to create and maintain a healthy environment.

Few persons are completely unaware of the dangers experienced by man in the technological world. But, on the whole, the public does not appear to be greatly alarmed about them. We tend to ignore progressive changes, until they reach catastrophic dimensions. For example, we barely notice that the air over urban areas is hardly ever luminous and fragrant. We have come to tolerate air pollution, foul water, crowding, garbage, noise, and confusion in the streets. Similarly, we shall probably learn to tolerate the sonic boom, invasion of privacy, and even the loss of freedom.

Adaptability is an asset for biological survival, but paradoxically, the greatest threat to the quality of human life is that the human species is so immensely adaptable that it can survive even under the most objectionable conditions.

There is much evidence that tolerance of undesirable conditions is achieved at the cost of physical and mental disabilities later in life. To a large extent, the so-called diseases of civilization are the delayed consequences of biological and mental stresses to which the organism has made adaptive responses that appear effective at the time they occur, but are inadequate in the long run.

The words soil, air, water, freedom are loaded with emotional content because they are associated with biological and mental needs that are woven in the fabric of man's nature. These needs are as vital today as they were in the distant past. Scientists and economists may learn a great deal about the intricacies of physicochemical phenomena, biological processes, and cost accounting. But scientific knowledge of environmental management will contribute little to health and happiness if it continues to neglect the human values symbolized by phrases such as the good earth, a brilliant sky, sparkling waters, a place of one's own. A true science of human life should concern itself with the maintenance and enlargement of the values that the eighteenth century associated with the word "civilization."

The role of science in the environmental crisis is ambiguous. Scientific technology has obviously produced a wealth of desirable effects which have been properly publicized; but it has also produced many undesirable ones which are unjustifiably soft-pedaled as "side effects." It would be intellectually more honest and socially more useful to publicize that practically all the advantages derived from scientific discoveries and technological achievements have to be paid for in the form of new dangers and new threats to human welfare. The fact that nuclear science promises endless sources of energy but also makes it possible to build ever more destructive weapons symbolizes the two faces of the scientific enterprise. All too often there exists a painful discrepancy between what man aims for and what he gets. He sprays pesticides to get rid of insects and weeds, but he thereby kills birds, fishes, and flowering trees. He drives long distances to find unspoiled nature, but he poisons the air and gets killed on the way. He builds machines to escape from physical work, but he becomes their slave and suffers boredom. The pages of newspapers and magazines bear witness to the public's somber anticipation that the legend of the sorcerer's apprentice may soon be converted from a literary symbol into a terrifying reality.

In the spring of 1968 many sheep died of a new kind of disease in valleys near the Army's Dugway Proving Grounds in Utah. (Brodine, 1969) Death was caused by minute amounts of a highly toxic organic phosphate that had been deposited by the wind on pastures over the mountain ranges while being tested as a biological weapon inside the Proving Grounds. Army scientists were naturally criticized for having used the poisonous substance before knowing how it moved through nature and before determining all its biological effects. The same criticism can be leveled against society as a whole.

Immense numbers of human beings are now being exposed from birth to a great variety of substances of unknown toxicity. It is probable that the new generations will suffer from this exposure more than adult persons, because they are being exposed early in life. If this is true the worst effects of pollution are yet to come. Since we make so little effort to investigate the effects of social and technological innovations on human life, we are practicing—not by intention but by irresponsible action—a kind of biological warfare against nature, ourselves, and especially against our descendants.

The total effect of the environmental crisis cannot be evaluated because it is spread throughout the whole social structure. Its economic costs might conceivably be measured, but there is no possible way to put a price on health, happiness, intellectual creativity, the esthetic appeal of scenery and buildings, the qualities of animal and plant life, and countless other values which are being altered by environmental influences. Yet, these values must be taken into account in any attempt to create and maintain a healthy environment.

The human organism is able to survive and function only within a fairly narrow range of physicochemical and social conditions; consequently, there are strict limits to the environmental changes that are biologically and socially safe. Furthermore, the phrase "healthy environment" implies the consideration of many factors beyond those having to do with biological and mental health, the conservation of natural resources, and the maintenance of ecologic equilibrium. As mentioned earlier, not only does man survive and function in his environment, he is *shaped* by it, biologically, mentally, and socially. To be healthy in the full sense of the word, the environment must provide conditions that favor the development of desirable human characteristics.

The very process of living involves a constant feedback between man and his environment with the result that both are constantly being modified in the course of this interplay. Individual persons and their social groups acquire their distinctive characteristics as a consequence of the responses they make to the total environment. The exciting richness of the human panorama results not only from the genetic diversity of mankind but perhaps even more from the fact that surroundings and ways of life shape the biological and social attributes of man.

Early in the twentieth century the American physiologist and sociologist L. J. Henderson developed the theme that the conditions prevailing on the Earth are uniquely suited to the emergence and maintenance of life. In his classical book *The Fitness of the Environment* (1913), he stated, "Darwinian fitness is compounded of a mutual relationship between the organism and the environment. Of this, fitness of environment is quite as essential a component as the fitness which arises in the process of organic evolution."

The word "fitness" means a high degree of

reciprocal adaptation between man and environment. More interestingly, it also implies that all aspects of human development reflect the adaptive responses made by the organism to environmental stimuli. In the long run, most forms of adaptation involve evolutionary alterations of the genetic endowment. But, in addition, the biological and mental characteristics of each individual person are shaped by his responses to the environmental forces that impinge on him in the course of his development. Genes do not determine the traits by which we know a person; what they do is govern his biological responses to environmental influences. As a result each person is shaped by his environment as much as by his genetic endowment.

The environmental influences that are experienced very early during the formative phases of development (prenatal as well as early postnatal) have the most profound and lasting effects. Early nutrition, education, technological forces, and esthetic and ethical attitudes are but a few among the multiple types of early influences that make an irreversible imprint on the human body and mind. Most of the biological and mental characteristics that are assumed to be distinctive of the various ethnic groups, anywhere in the world, turn out to be the consequences of early environmental influences (biological and social) rather than of genetic constitution. (Dubos, 1968[b])

Mental development is stimulated if the child is exposed at a critical time to the proper range of stimuli and acquires a wide awareness of the cosmos. Science and technology can play a crucial role in the shaping of mental attributes by making it possible to create environments more diversified and thereby more favorable for the expression of a wider range of human potentialities.

All men are migrants from a common origin. They have undergone biological and social changes that have enabled them to adapt to the different conditions they have encountered in the course of their migrations. But as far as can be judged, all ethnic groups are similarly endowed with regard to biological and mental potentialities. This fact justifies the belief that, given the proper opportunities, any population can shape its future and select the form it gives to its own culture by focusing its attention on the biological, technological, and social forces that affect human development.

The Spaceship Earth

In his last speech, delivered before the United Nation's Economic and Social Council in Geneva (1965), Ambassador Adlai Stevenson poignantly expressed man's dependence on the conditions prevailing over the whole earth: "We travel together, passengers on a little spaceship, dependent on its vulnerable supplies of air and soil; all committed for our safety to its security and peace, preserved from annihilation only by the care, the work, and I will say, the love we give our fragile craft."

Despite Stevenson's warning human ecology is still a no-man's land. Physicians and biologists study in great detail bodily structures and functions. Technologists develop ingenious and powerful techniques for manipulating nature and creating artifacts. But little attention is paid to the interplay between man and the world in which he lives. Yet, it is obvious that human biology should be primarily concerned with the responses that the body and the mind make to surroundings and ways of life. Ecological problems are difficult to deal with because we lack methods for investigating scientifically the interrelatedness of things.

The expression "Spaceship Earth," as used by Stevenson, is no mere catch phrase. Not only do we live on earth, but we evolved on it; as a result, our biological and mental being has been shaped by the conditions peculiar to its crust. We cannot survive without continuously drawing breath from its shallow layer of air, using and re-using its limited supply of water and other essential resources. Yet, we proceed as if we were not aware of its limitations and of the threats to its safety and to our own inherent in the modern ways of life.

It would be easy, far too easy, to conclude from present trends that mankind is on a course of self-destruction, even if a portion of it escapes nuclear warfare, mass hunger, wholesale poisoning, or the disasters likely to result from excessive population density. In any case, the quality of life will inevitably deteriorate as the closed environment of the Spaceship Earth continues to become more crowded, depleted, polluted, and desecrated. We shall progressively lose the social attributes identified with civilized humanness if very soon we do not stop population growth and redirect technology into a more sensible course.

This is not the first time that men have been crowded and have spoiled their surroundings. In the past, however, they could move to some other parts of the planet and establish new settlements. Such migration is now all but impossible, because practically all habitable parts of the globe are occupied and most of them are already becoming polluted. Nor is there any chance that we can escape into other worlds. As already mentioned, the range of conditions under which he can survive and function is extremely narrow and exists only on the surface of the earth. Despite the irresponsible assertions of a few scientists and the imaginings of science fiction writers, the world population is therefore bound to the earth by the exigencies of man's biological nature.

Since the birth rate now greatly exceeds the death rate almost everywhere, the world population is soaring. This situation is not peculiar to the underdeveloped parts of the world; it is equally as threatening in North America and Western Europe as in South America and Asia. Even though the birthrate has been falling recently in industrialized countries, the fall is not occurring fast enough to prevent the populations from increasing rapidly in most of these countries. In the United States the increase currently is approximately 2,500,000 persons per year, and a similar rate of growth is likely to continue for at least a few decades. This added population would require a new large city or ten medium–size cities every single year, if we want to avoid further growth of the present urban agglomerations.

Furthermore, economic and social forces are driving people from the country to urban areas, and this trend also is likely to continue. Under the present conditions of life, high population density always results in gross environmental pollution by man's own biological processes and even more by technology. No one knows exactly the maximum size of the population that can safely live on Earth; but it certainly cannot be many times greater than what it is at present. Technologists may develop nuclear energy, labor saving gadgets, new synthetic chemicals, scientific agriculture, ready-made housing development, and cities of extravagant height and length but the overpowering fact is that we shall soon be running out of desirable places and essential resources. As Thoreau remarked, "What is the use of a house if you haven't got a tolerable planet to put it on?"

The spectacular achievements of technology during the past few decades and the confidence that most aspects of the physical world and of man's nature will eventually be understood and brought under control would seem to warrant the hope that scientific solutions can be found for the problems now facing mankind. Several groups of scholars, as we have seen, recently published forecasts of the technological and medical advances to be expected for the twenty-first century (pages 59-61).

In the last chapter of his book *The Most Probable*

World (1968), the American social critic Stuart Chase describes the ideal suburban life under unpolluted skies along sparkling streams that will be possible after the present problems of the technological environment have been solved by scientific management. However, Chase devotes only four pages of his book to his suburban utopia as against 226 pages devoted to the problems that presently threaten to destroy technological civilization. This proportion symbolizes the discrepancy between the hypothetical world of scientific scenarios and the actual technological environment. The rational efficient, silent, and luminous city of the future forecast in the General Motors' Futurama at the New York World's Fair of 1940 has become, in fact, the hellish agglomeration of the 1960's, more confusing and traumatic as it becomes larger. Despite the unquestionable possibility of scientific and technological breakthroughs in the future, nothing of significance is being done to deal with the problems that threaten mankind today.

The advances discussed in books concerned with the future of scientific technology are usually spectacular in character and occasionally of theoretical interest. But none of them are designed to improve the quality of human life. The authors of *The Year 2000* and *The Year 2018* do not discuss, for example, the rape of nature, environmental pollution, the feeling of anomie, racial conflicts, and other problems which are on everybody's mind. The breakthroughs emphasized in these books relate to the production of power rather than to the understanding of its use by people;

the books deal primarily with inanimate things and almost ignore life. Far from helping in the solution of environmental problems, such new breakthroughs will certainly create new dangers and accelerate the degradation of nature.

The man of flesh and bone is not likely to remain long impressed by the fact that a few of his contemporaries can explore the moon, program their dreams, or use robots as slaves, if the planet earth becomes unfit for his everyday life. He may soon tire of space acrobatics if his eyes smart from smog and his feet are deep in garbage.

For the past three centuries Western civilization has been largely based on Francis Bacon's motto that "knowledge is power." It is now apparent, however, that using the power produced by scientific technology for mastering nature is not sufficient to sustain civilization, not even industrial civilization. Our societies are beginning to realize that the very social and technological practices that have made them economically prosperous and politically powerful, increasingly damage human and environmental health. Slowly and grudingly they are developing palliative measures to deal with the most obvious threats; but this piecemeal approach will not be sufficient to solve the ecological crisis and improve the quality of life. Technological fixes amount to little more than putting a finger in the bursting dike, whereas what is needed is a sociotechnological philosophy of man in his environment.

Political reasons have stimulated the study of a few ecological problems involving complex relationships between man and his environment, for example: planning river basins for water, land, and industrial management, as in TVA; training combat forces for operation in the arctic or the antarctic; and the preparation of astronauts or aquanauts for travel or residence in space or underseas. But the knowledge thus acquired is so highly episodic that it does not allow for theoretical generalizations applicable to many other systems. Instead of carefully working out long-range programs of action based on ecological understanding, the tendency has been to develop ad hoc countertechnologies for the piecemeal correction of the new problems created by technology. In the words of the American social critic Gerald Sykes, "Man rushes first to be saved *by* technology, and then to be saved *from* it. We Americans are front runners in both races."

Countertechnologies, also called technological fixes, are at best short-range palliatives which usually create new environmental problems of their own. Superhighways and mammoth underground or multistoried garages constitute almost caricatures of countertechnologies; they are designed to facilitate traffic, but in fact they encourage the proliferation of automobiles, thereby increasing congestion and pollution in the cities. The association between monocultures and the use of pesticides also illustrates the kinds of problems commonly created by countertechnologies. Monocultures (of almost any crop) make possible the use of farming practices which increase yields and decrease production costs; but they also encourage the development of plant pests which require the use of more and more pesticides; beyond a certain point pesticides become a serious threat to human health, and this in turn promotes control measures which become increasingly annoying and costly as the monoculture technology becomes more widespread.

Many medical problems characteristic of our times can be similarly traced to the unpredicted consequences of medical technologies and countertechnologies. A simple example is the concatenation linking cortisone to metabolic and cellular changes that increase susceptibility to infectious disease. Antibacterial drugs may control the bacterial infections evoked into activity by cortisone, but they also open the way for invasion by yeasts and fungi that in turn demand the use of still other drugs. Kidney dialysis, organ transplantation, and the various forms of prostheses are already beginning to create their own medical, social, and ethical problems. One can anticipate that the wholesale use of contraceptives and the systematic limitation of family size will also create difficulties which may even involve man's genetic endowment. With the present death rates, population size can be stabilized only if family size is limited to 2.2 children per couple. It is not unlikely that such drastic limitation will have genetic, physiological, and emotional consequences that will prove unfavorable in the long run.

Technological fixes are of course needed to alleviate critical situations, but generally they have only temporary usefulness. More lasting solutions must be based on ecological knowledge of the physicochemical and biological factors that maintain the human organism in a viable relationship with the environment. In most cases such knowledge is not available because certain areas of science have not yet been sufficiently developed or have been neglected altogether. Nature conservation and human adaptation constitute important examples of fields that deserve far greater emphasis than they are receiving from the scientific community.

Many technological and social developments endanger the supply of natural resources and thereby the quality of human life. The word conservation encompasses the various modalities of this problem and therefore means different things to different persons, for example: (a) preserving as much of nature as possible in its primeval state; (b) maintaining unspoiled certain aspects of the natural and civilized world that we consider particularly enjoyable and desirable, such as the wild life of marshlands, the beautifully humanized fields and barns of the Pennsylvania Dutch country, the temples, cathedrals, palaces, and homes that testify to the glory of human life in the past; (c) manipulating the environment in such a manner that it evolves in harmonious relationships with human beings, whether these elect to live in isolated cottages or in skyscrapers, whether they favor a pastoral way of life or an electronic civilization; (d) planning environmental utilization in such a manner that it is compatible with economical use of natural resources and with their renewal.

Each of these different philosophies of conservation has its merits and corresponds to worthwhile social goals. But the philosophies can be viable, and their goals reached, only if certain ecological imperatives are respected. The formulation of these imperatives demands a kind of scientific knowledge far different from what is taught in schools and from what present research institutes are developing.

It is impossible at present to plan for the systematic teaching of this knowledge, since it exists only in rudimentary form; but it would be possible to create institutions with an intellectual atmosphere and research facilities favorable for its development. Environmental values are so intimately and inextricably enmeshed with human values that a new kind of ecological science will be needed to encourage conservation to evolve from a sentimental attitude to an operational concept.

At first sight the problems of human health appear different from, and more precisely defined than, those relating to environmental health. Yet, there are analogies between the two states. Like environmental control, the science and practice of medicine have been traditionally concerned more with palliative measures such as the treatment of disease than with the formulation of long-range policies aimed at the maintenance of health.

One can anticipate the discovery of procedures for alleviating most human ills—drugs to correct physiological disorders or arrest infections; techniques for the removal, correction, and transplantation of organs; development of mechanical prostheses for

replacing diseased parts; and procedures for manipulating beliefs and moods. But it will probably be impossible in the foreseeable future to make the best methods of treatment available to all persons in need of them. No society, however prosperous and generous, will be able to carry the economic load, and especially provide the large numbers of highly-trained personnel, that would be required to do all that could be and should be done.

On the other hand epidemiological evidence leaves no doubt that many of the chronic and degenerative disorders which constitute the most difficult and costly medical problems of our societies have their origin in the surroundings and in the ways of life rather than in the genetic constitution of the patient. But little is known of these environmental determinants of disease.

Medical schools and institutes of biomedical research elected half a century ago to focus their interest on the detailed aspects of structure and function—the result being that little heed has been paid to the responses that progressively shape the organism in the course of its development. Yet, it is certain that the phrase "human constitution" implies much more than the genetic endowment. The characteristics of a person and the responses (healthy or pathological) he makes to the environment are profoundly conditioned by the past experiences that he has embodied in his biological and mental being. Human constitution can thus be defined as the continuously evolving phenotype—i.e., the incarnation at any given time of a particular person's experiential past. (Dubos, 1959, 1965, 1968[b])

The change in phenotype might have been called "allergy," but unfortunately the meaning of this word is now restricted to a limited range of pathological conditions. The etymological meaning of allergy is of course "altered reactivity." In this sense many aspects of life can be considered as manifestations of allergy, since the anatomical and functional characteristics of man can be modified by the stimuli—physical, chemical, and social—that impinge on him throughout life. These effects are particularly profound when they occur during the early, formative phases of development.

The range of man's potentialities is largely uncharted, and we are also very much in the dark concerning his developmental responses to stimuli. But enough is known to establish that the environmental forces that impinge on him during the early, critical stages of development influence profoundly, lastingly, and in certain cases irreversibly his most important biological and mental characteristics, such as initial rate of growth; age of sexual maturity; maximal size of the adult; numbers of cells laid down in muscular, adipose, and brain tissue; learning ability; and behavioral patterns and emotional attitudes. As used in the preceding sentence, the phrase "critical stages" denotes the various periods of prenatal and postnatal life during which the phenotypic attributes that define biological and mental individuality achieve their characteristic organization.

Before discussing further the effects that the environment exerts on man, it seems useful to emphasize that the essential genetic aspects of his body and brain were shaped by the environmental forces that prevailed on earth during his evolutionary past. Furthermore, as far as can be judged, man's genetic makeup has not changed significantly since the Old Stone Age. In the words of the English anthropologist Jacquetta Hawkes (1968), "The psyche which lives in every human being as the generations rise and fall has not been totally transformed. It makes an unbroken chain between present and past. The bodily experiences opened to the men and women of thousands of years ago were identical with those of today; the mental and emotional ones were not altogether different."

The modern environment naturally exposes man to conditions very different from those under which he evolved. To a certain extent genetic readaptation to new environmental forces can take place because the huge range of potentialities that have accumulated in the human gene pool during the evolutionary past can be redistributed and selected to produce a rapid genetic drift whenever the selection pressure is strong enough. But, wide as man's variability may be, it is limited by the instructions encoded in his genetic equipment. Scientific civilization must therefore concern itself with the genetic limits of man's tolerance, because these determine the range within which social and technological changes are compatible with his life.

Immunity and allergy (in the usual sense of the

word) are the two best-studied types of lasting changes that the environment can elicit in man. But they are not the only ones, nor even the most important. As mentioned earlier, most aspects of the body and the mind can be altered almost irreversibly by the organism's responses to almost any stimulus. For this reason the quality of the environment cannot be judged only from its present effects; the delayed and indirect effects may be more important in the long run.

It has been shown, for example, that a single injection into newborn mice of particulate materials separated from urban air greatly increases the frequency of various types of tumors during the adult life of these animals. If this observation can be extrapolated and applied to human beings, the worst consequences of pollution are yet to be recognized, since it is only during the past two decades that babies born and raised in urban areas have been exposed to high levels of pollutants.

If present trends continue most people in the world will soon live in large urban agglomerations. Massive urbanization would have disastrous biological consequences if it were not for the fact that the majority of urban dwellers develop some form of tolerance to environmental pollutants, intense sensory stimuli, and high population density, just as they develop herd immunity to ubiquitous microbial pathogens. However, such acquired tolerance may not be an unmixed blessing, because it is often the expression of undesirable functional changes. For example, continuous exposure to low levels of air pollutants stimulates mucous secretion in the walls of the tracheobronchial tree, thus affording some protection to the pulmonary epithelium. But eventually the cumulative effects of irritation result in chronic bronchitis and other forms of irreversible pulmonary damage.

Recent physiological and behavioral studies have revealed that people born and raised in an environment where food intake is quantitatively or qualitatively inadequate achieve a certain form of physiologic and behavioral adaptation to low-food intake. They tend to restrict their physical and mental activity and thereby reduce their nutritional needs; in other words, they become adjusted to undernutrition by living less intensely. Furthermore, they retain throughout their lives the physiological and mental imprinting caused by early nutritional deprivation. Physical apathy and indolence have long been assumed to have a racial or climatic origin. In reality, these behavioral traits often constitute a form of physiologic adjustment to malnutrition, especially when nutritional scarcity has been experienced very early in life.

Undernutrition is now rare in affluent countries, but malnutrition can take many other forms, including perhaps excessive artificial feeding of the infant. Little is known of the biological and psychological effects that result from a nutritional regimen which differs qualitatively from that of the mother's milk and exceeds it quantitatively, but there is evidence that infants fed an extremely rich and abundant diet tend to become large eaters as adults. Such acquired dietary habits are probably objectionable from the physiological point of view; it would be surprising if they did not have behavioral manifestations.

Crowding and constant exposure to social stimuli are inevitable accompaniments of urbanization and are usually regarded as deleterious. Human beings, however, can become adapted to crowding, especially if they have been exposed to it during the early phases of their development—just as they can become tolerant to most other types of stressful situations. Adaptation to crowding has been observed in laboratory animals and in human populations.

Experiments in various animal species have revealed that crowding commonly results in disturbances in endocrine function and behavior. But the intensity of the effects is profoundly influenced by the conditions under which high population density is achieved. When adult animals of a given species are brought together in a confined environment, they exhibit aggressive behavior and a large percentage of them may die. In contrast, if a few animals are placed in a given enclosure and allowed to multiply in it, the population to which they give rise can reach very high density without evidence of destructive aggressiveness. While growing together the animals achieve a social organization that minimizes violent conflict. Beyond a certain level of population density more and more animals exhibit abnormal behavior, but in general these deviants are not sick organically. They act as if they were unaware of the presence of their cage mates; their behavior is asocial rather than antisocial.

Men are not rats, yet the most unpleasant thing

about overcrowded rats is that they behave so much like human beings in some crowded communities. Man has developed a variety of social mechanisms that enable him to live at high-population densities; for example, Hong Kong and Holland show us that such densities are compatible with physical health and low crime rates. However, there are other human communities in which extreme crowding leads to a kind of asocial behavior very similar to the social unawareness manifested in overcrowded animal populations.

In most cases the deleterious effects of crowding result not so much from high population density as from the social disturbances associated with *sudden* increases in density. The appalling amount of physical and mental disease during the first phase of the Industrial Revolution was caused in large part by poor sanitation and malnutrition, but certainly another important factor was that immense numbers of people from rural areas migrated within a few decades to the new industrial cities. They had to live and function in the crowded slums and shops of the teeming industrial cities before they could make physiological and emotional adjustments. Yet it took but a few decades to convert these rural populations into urban ones, which now find satisfaction in the crowds that precipitated biological and mental diseases a century ago.

Because of my professional specialization and also probably because our society is disease-conscious, the pathological aspects of response to the environment have been emphasized; but there are other psychological aspects which are at least as interesting and socially important.

In the modern industrial city human beings hardly ever have the chance to see the Milky Way or a night radiant with stars or even a truly blue sky. They never experience the subtle fragrances peculiar to each season, the exhilaration of early spring and the poetic melancholy of autumn. The loss of these experiences may be more than an esthetic deprivation; certain emotional needs were woven in man's fabric during his evolutionary past, and their satisfaction may well be required for complete biological and mental sanity.

Other kinds of undesirable changes are likely to occur as a result of extreme urbanization. The complexity of social structures makes some form of regimentation unavoidable; freedom and privacy may come to constitute antisocial luxuries, and even to involve hardships. As a consequence the human beings most likely to prosper in congested urban environments will be those willing to accept a regimented life in a teeming world from which all wilderness and fantasy will have disappeared. The domesticated farm animals and the laboratory rodents on controlled nutritional regimens in controlled environments will then become true models for the study of man.

Admittedly, it is possible to rear and train children for oversocialized conditions—to such an extent that they do not feel safe and happy outside a crowd of their own kind. But this does not invalidate the view that there is potential danger in increased urban crowding. Children and even adults can be trained to accept as desirable almost any form of perversion—physiological, behavioral, or intellectual.

In the final analysis, the human environment is what man experiences, and it is the quality of this experience that shapes individuality and gives its value to life. For this reason one feels sorrow and indignation at seeing the children in American cities being continuously exposed to noise, ugliness, and garbage in the street—and thereby conditioned to accept public squalor as the normal state of affairs.

Winston Churchill expressed in a memorable phrase that the environment has a deep and lasting influence on human beings: "We shape our buildings and afterwards, our buildings shape us." In even more telling and moving words, James Baldwin has repeatedly affirmed his conviction that the American Negro is shaped by his life experiences, especially by the early ones. The following passages convey what it means to have grown up as a Negro in the United States (1963):

> We cannot escape our origins, however hard we try, those origins which contain the key—could we but find it—to all that we later become.
>
> It means something to live where one sees space and sky or to live where one sees nothing but rubble or nothing but high buildings
>
> We take our shape . . . within and against that cage of reality bequeathed us at our birth.

The Spaceship Earth is the cage within and against

which man has developed in his evolutionary past and continues to develop his biological and mental characteristics. As the terrestrial environment deteriorates so does humanness and the quality of human life.

Admittedly, human beings are so adaptable that they can survive, function, and multiply despite malnutrition, environmental pollution, excessive sensory stimuli, ugliness, and boredom, high population density and its attendant regimentation. But while biological adaptability is an asset for the survival of *Homo sapiens* considered as a biological species, it can be fatal to the attributes that make human life different from animal life. From the human point of view environmental quality and the success of adaptation must be judged in terms of values peculiar to man.

A recent issue of the *Medical Tribune* (1967) carried an essay with the surprising title "Villagers *Adapting* to their Arsenic-filled Water" (italics mine). According to the author of this essay the underground water used by the inhabitants of the Mexican village Finisterre, in the province of Cocihuila, is heavily contaminated with arsenic (628-949 micrograms per liter). As a result, more than two-thirds of the villagers exhibit severe symptoms of chronic poisoning, such as neurological disorders, blood protein abnormalities, goiter, and skin lesions. But most persons so affected are nevertheless able to work. In the words of the *Medical Tribune*, they "appeared well *accustomed* to their disorder" (italics mine), even though obviously suffering from chronic arsenic poisoning.

The words "adapting" and "accustomed" have

ambiguous meanings. While most inhabitants of Finisterre continue to function, they are obviously sick. What the writer had in mind when he put the word "adapting" in his title was that the Mexican villagers have come to accept as inescapable facts of life the poor quality of their water supply and its pathological consequences.

We have little arsenic poisoning in the United States, but like the Mexican villagers we are "adapting" to environmental insults that spoil the quality of life. We also do little if anything to prevent or correct these conditions, and we behave as if they were inescapable acts of God. There are many historical precedents for such passive acceptance of environmental threats. For example, Jacob Bigelow wrote in his *Modern Inquiries* more than 100 years ago: "Should the cholera continue to prevail for three years throughout this continent, it would cease to interrupt either business or recreation. Mankind cannot always stand aghast; and the wheels of society at length would be no more impeded by its presence than they now are by the existence of consumption, of old age, or drunkenness."

Our own failure to prevent environmental degradation cannot be accounted for by lack of awareness. We would like to improve our polluted and cluttered environments, but we like gadgets and economic prosperity even more. In fact, values such as political power and gross national income so dominate our collective lives that we shall undertake the social and technological reforms essential for environmental control only if we are forced into action by some disaster. That environmental disasters will occur seems to me inevitable.

A prolonged period of thermal inversion caused the smog that killed many persons in the small town of Donora, Pennsylvania in 1948. The numbers of fatalities will be magnified many thousandfold when inversion lasts a few days longer than usual over any one of our large cities; we were close to this situation in New York during the Thanksgiving week of 1966. The thalidomide episode has made painfully evident the fact that an apparently innocuous substance can have disastrous effects before its toxicity is recognized; it would be surprising if one of the countless new substances that now reach the whole population through air, water, and food did not turn out to have deleterious effects. And when this happens millions of persons will be affected almost simultaneously.

Direct health dangers are not the only distressing aspects of the environmental crisis. Inorganic and organic wastes are accumulating so rapidly that their management is now at least as important as the production of new resources for the survival of civilization. Wastes will soon convert the biosphere into a global dump unless they are dealt with at their source, before they become a public nuisance. Also it is important to recover wastes before they are dispersed through air, water, and soil, because resource production now depends upon the utilization of wastes. Converting wastes into usable products can thus contribute to environmental quality both by minimizing pollution and retarding the exhaustion of natural resources. (Restoring, et al., 1965).

The enactment of policies for environmental improvement will certainly create conflicts with individual interests, but many of these conflicts could probably be resolved if scientists and technologists were to focus their efforts on the problem.

When the Leblanc process for soda ash manufacture was introduced in England around 1830, extensive environmental damage was caused by the hydrogen chloride gas emitted from the factories. In 1863 Parliament passed the Alkali Act, which called for a 95 per cent reduction in hydrogen chloride emissions. Almost immediately technologists developed better gas absorbers and commercial processes for converting hydrogen chloride into chlorine gas. Half a century ago the dairy industry opposed legislation for pasteurization of milk on the grounds that it was impractical and would be too costly. But once the law was enacted scientific technology developed a safe, efficient, and inexpensive method of pasteurization. Instead of pricing milk out of the market, as had been predicted by the dairy industry, pasteurization decreased its price by increasing its shelf life.

Noise control provides a timely illustration of the fact that both the social and the natural sciences have an important role to play in making technology compatible with environmental quality. Since steady intense noise can inflict irreversible damage, American courts award compensation to workers whose hearing has been affected by exposure during work. Yet, many urban dwellers are constantly exposed to levels of noise that often equal and may exceed those considered on the borderline of safety in industrial work. Furthermore, recent laboratory investigations indicate that supposedly tolerable noise levels can cause ear damage in animals and that sounds not sufficiently loud to awaken sleeping persons nevertheless affect their brain waves. Most disturbing in this regard is the recent discovery that the heart rate of the fetus can be accelerated by noises to which the mother appears to have become tolerant. One shudders at the thought of what the supersonic boom will do to the forthcoming generations if it ever becomes a part of daily life—not to mention a variety of unpredictable accidents such as those caused by the jarring of a surgeon's hands during an operation. People may learn to tolerate the boom, but the chances are very great that this adjustment will be achieved at the cost of physiological and mental damage. In many cases the sonic boom may turn out to be a sonic doom.

Physicists seem to agree that there is no way to prevent the boom at supersonic speeds or to decrease its effects significantly by high altitude flying. In this case, therefore, scientific knowledge can contribute to environmental quality by making clear where a technological solution *cannot* be found. Furthermore, sociological knowledge shows that the man-hour loss of time by persons experiencing the boom and recovering from its effects far exceeds the man-hour saving of time by persons flying at supersonic speed. The solution of rapid air transport must therefore be sought in directions other than the improvement of supersonic aircraft, for example, in more efficient ground operations in and out of airports.

The noise made by more conventional equipment presents a different kind of problem which seems amenable to technological solutions. Buses, trucks, subways, elevated trains, planes, helicopters, air conditioners, office machines, construction and sanitation equipment—all these and many other sources of high-level noise can be made quieter through proper engineering techniques. Acoustical science and technology have provided the armed forces with an inaudible motor for front-line use, silently operating submarines, and almost silent airplanes. Surely some of these same techniques could be applied to civilian use and thus minimize the health hazards of excessive noise.

Really effective methods of environmental control will of course be costly. Congressional committees which have begun calculating the cost of providing clean air and water have come out with estimates of several hundred billion dollars! These figures are frightening even for a prosperous economy, but they do not compare with the staggering damage being done to man and his environment every year and in increasing amounts. So far we have disregarded this human and environmental damage in order to facilitate easier and more profitable technology, even if the end is the production of dangerous or useless gadgets or substances. This attitude is no longer permissible or possible. Environmental threats are becoming more dangerous at an accelerated rate, in part because technology continuously increases the complexity of life and also because the space available to each person decreases as the population increases. One person's trash basket is another's living space, and everybody lives within hearing distance of everybody else. As the earth becomes more crowded, there is no longer an "away."

Developing *countertechnologies* to correct the new kinds of damage constantly being created by technological innovations is a policy of despair. If we follow this course we shall increasingly behave like hunted creatures, fleeing from one protective device to another, each more costly, more complex, and more undependable than the one before; we shall be concerned chiefly with sheltering ourselves from environmental dangers while sacrificing the values that make life worth living.

The present ecological crisis throughout the technological world presents striking analogies with the human and environmental problems created by the first phase of the Industrial Revolution in Western Europe and in the United States. One historical precedent may provide the pattern for what we can expect and hope in the near future.

A century ago several outbreaks of cholera created an atmosphere of near panic in several large European and American cities. The outbreaks were limited in scope but sufficiently alarming nevertheless to make the public aware of the horrifying conditions prevailing in most urban areas and of the urgent need for environmental reforms and changes in the ways of life. As a result, a vigorous campaign was launched along three different approaches:

(1) enlightened and dedicated laymen organized programs for cleaning up the urban mess and for assuring to every citizen "pure air, pure water, and pure food"; to these demands we should add today quiet and privacy;

(2) civil servants supported by lay groups established boards of health for the formulation and enforcement of sanitary regulations, often against the resistance of vested interests; and

(3) scientists established new research institutes to study the epidemic diseases which constituted the most pressing environmental problems of the period.

These three different approaches can serve as a pattern for social and scientific action in our own times. Environmental degradation in all its forms is everybody's business; its control will require a massive mobilization of public, administrative, and scientific concern.

In many cases, control will conflict with individual interests, but this should not serve as an excuse for failure to act. We can trust that socially acceptable technological solutions can be found if the scientific community faces up to the problems posed by the environmental crisis.

Modern scientists give much lip service to their social responsibilities, but in practice they behave as if they were captives of an establishment which often appears asocial and not infrequently antisocial. If a massive effort similar to the one represented by the National Aeronautic and Space Administration (NASA) is not soon initiated to deal with the environmental crisis, then we will have to conclude that the scientific community and the governmental agencies responsible for the funding and administration of science are not as concerned with human welfare as they pretend to be. A similar situation prevailed at the beginning of the nineteenth century. Public pressure, organized by enlightened laymen, was then the force that eventually placed environmental problems at the forefront of the scientific endeavor. Once more, a similar grass roots movement is urgently needed to convince public bodies and the scientific establishment that high priority should be given to the study and control of the forces that affect the quality of human life and its environment and that are rapidly making the Spaceship Earth a place unfit for human life.

The fundamental unity of the human mind gives a universal quality to all forms of knowledge, especially to theoretical scientific knowledge. But while the facts and laws of science are valid everywhere, the applications of science differ from one area and from one group of people to another. The problems of India cannot be solved by using the practical knowledge and the technologies developed in Indiana. A few examples will suffice to illustrate the need to consider local and regional characteristics in the development of scientific knowledge.

The technical problems of agriculture and conservation are different in arid, temperate, and tropical areas. Soil management, animal husbandry, and plant rotations must be designed in the light of the geologic, climatic, and social conditions peculiar to each locality.

Malnutrition may be due to shortage of calories in certain parts of the world, to vitamin deficiencies in others, and to shortage of good quality protein elsewhere. The development of complete food preparations that can serve as substitutes for animal and dairy products must be based on the kind of plant resources that can be economically produced, and this in turn depends upon local geologic and climatic conditions.

A recent UNESCO conference (1968) urged the establishment of programs for monitoring pollutants in entire air sheds and water basins; but pollutants differ from region to region. Air pollution on the Pacific Coast of the United States differs chemically from what it is in Northern Europe or in Taiwan. Water and food are usually polluted with microbes in certain parts of the world but not so much in industrialized countries where chemical pollution may be more important.

Cosmic rays at high altitude, radio-nuclides absorption in areas of high radioactive background, and marine chemistry and biology on different types of shore lines are but a few of the many examples that may have great potential importance for different countries in a same region.

On the other hand, there are scientific problems

that obviously involve the whole world community. For example:

—Weather modification—who will be deprived of water if rain is made to fall on a given area?

—Control of epidemics—how fast and along what routes do the various strains of influenza virus spread from one continent to another?

—The protection of endangered species—certain species of primates are used on such an enormous scale in American and European laboratories that steps should be taken to protect their populations in the countries of origin;

—Brain drain and related problems pertaining to the education and utilization of scientists.

Whether primarily concerned with man, his ways of life, or his surroundings, most scientific problems which are socially important require research facilities that few institutions or even nations can afford—hence, the need for a cooperative approach among mission-oriented institutions, either at a regional or global level. Fortunately, the experience of the past few decades indicates that supranational scientific centers can be effective.

The World Health Organization and its multifarious control and study programs—the International Geophysical Year—and the World Meterological Organization and its plans for a World Weather Watch all illustrate that scientific research and action on a global scale can be highly successful. The Institute of

Nutrition for Central America and Panama (INCAP) in Guatemala City and the Satocholera Laboratory in East Pakistan can serve as examples of regional institutions devoted to problems of health. The Centre European pour la Recherche Nucleaire (CERN) in Geneva, Switzerland, and the International Center for Theoretical Physics at Trieste, Italy, illustrate the potentialities of international laboratories for theoretical science. The success of these very different types of international institutions should encourage the development of other research programs on a regional or global scale.

In the speech mentioned at the beginning of this chapter, Adlai Stevenson spoke of the care, the work, and the love we must give to the Spaceship Earth; but he also pleaded for more scientific knowledge of its resources and needs. So far, the management of our planet has been largely a haphazard enterprise, limited in most cases to the application of palliative measures for the containment of dangerous or objectionable conditions. Now that science has made us so destructive, because so powerful, we must try to imagine the kind of surroundings and life we want, lest we end up with a jumble of technologies that will eventually smother body and soul.

Science Without Jargon

The modern edifice of professional science is inevitably a tower of Babel. As knowledge develops specialists in any field develop a jargon which progressively becomes more incomprehensible to the general public and to specialists in other fields. This loss of ability to communicate does not occur only in the natural sciences. Experts in philosophy or economics find it just as difficult to communicate their professional knowledge to laymen or to other scholars as do mathematicians, physicists, or geneticists.

We must reconcile ourselves to the fact that the world is made up of an immense variety of specialized groups, each with its own body of experiences, words, and symbols. A few scientists, however, still manage to convey the essence of their specialty to other human beings by formulating their knowledge at a level of discourse that transcends their professional jargon. I shall first consider a few principles bearing on the communication of knowledge among scientists, then discuss what aspects of this knowledge can be usefully conveyed to laymen, keeping especially in mind the social implications of science.

It is often stated that the era of scientific books and journals is coming to an end because knowledge can now be electronically stored as separated bits of information and retrieved at will whenever needed. There is more to knowledge, however, than the ability to retrieve from a storage system the fragments of information that are needed for a particular purpose at a particular time.

Children become aware of the external world not as isolated bits of reality but as total situations; they learn language not as individual words but as structures of meaning. Adults probably acquire most of their knowledge through a similar learning process. They assimilate isolated facts by apprehending them in the form of integrated patterns.

In preliterate societies communication occurred largely through symbols. Knowledge was communicated by attitudes, gestures, images, or artifacts, and perhaps more generally by oral transmission of poetry, songs, traditions, or myths. Therefore such symbolic and oral transmission of knowledge depended on patterns and structures rather than on precise descriptions of specific objects or events. But it was nevertheless amazingly effective and faithful to fundamental themes, and at times it was even accurate regarding factual details.

The invention of writing and printing has not radically changed the fundamental processes through which information is conveyed; it has merely provided more effective and more convenient technologies for these processes. The most influential and successful books are still those which present facts and thoughts as organized patterns. The great scientific books are not necessarily the ones that are most accurate, richest in details, or best written but rather those with an interesting structure. The character of a book—in contrast to the amorphous nonbooks—reflects the personality and reflections of the author.

Etymologically, the word "information" denotes a process that gives *form* to knowledge, and therefore implies much more than the mere transmission of fragments of knowledge. The etymological root of information is *form*, which suggests that the fragments of knowledge must be organized into a structure before it becomes meaningful.

In the long run, communication probably is most productive and useful when it evokes a creative response on the part of the recipient. At its best, information is *formative*. What becomes incorporated in the reader is seldom exactly what the author thought and wrote. The reader uses the pattern provided by the author to create from it another structure influenced by his own *Weltanschauung*. As the reader evolves with time, the new structures he creates out of the pattern provided by the author likewise evolve. Reading is a formative activity during which the reader becomes involved with the author in a process akin to Buber's "I-Thou" relationship—the result being a new creation integrating some characteristics of both.

True information thus always means the communication of a structure and ideally the facilitation of a formative process in the recipient. Many different media can be used to this end, and indeed many have been used since the beginning of time. The drawings and paintings of prehistoric men, the poems and legends transmitted orally during the preliterate period, and the inscriptions on clay or papyrus were means of communication which proved immensely effective. Now books, newspapers, magazines, radios, neon signs, and television channels all play their part in the process that has, since the Stone Age, continuously *formed* the mind of man through communication of structured knowledge. The role of the communicator is to organize the fragments of knowledge into the multifarious patterns and structures through which man apprehends reality. The message is not in the medium but in the *structure* conveyed by the medium.

The need to convert fragments of knowledge into structured information is greatest with regard to science writing for the general public. Such writing can be considered from several points of view: (a) communication aimed toward the public understanding of scientific facts and laws—irrespective of their relevance to practical problems; (b) the communication of digested knowledge to be used by the public itself to some practical end—for example, health practices and other forms of "how to" information (in this case, information is essentially a kind of instruction; and, (c) recognition and evaluation of the consequences of science, especially of scientific technology and biomedicine, as they affect human life and social institutions.

Whatever its ultimate purpose, the presentation of science to the general public must go beyond describing factual knowledge in a simplified form. Knowledge must be structured with a particular purpose in view. Basically, the needs of the general public are the same as those of the student, or even of the research scholar, the difference being chiefly in the technical level of the language.

Ignorance of the law, the saying goes, is no excuse. In technicized societies some knowledge of science is as essential as knowledge of the law, because all social decisions now have scientific determinants. This does not mean that everyone should be a science graduate, anymore than it means everyone should have a law degree. The kind of scientific knowledge needed by the citizen is not the technical knowledge of the professional scientist, but a general understanding that will help him recognize, evaluate, and to some extent anticipate the social consequences of science and technology. For lack of this understanding, human beings increasingly will have to submit to the tyranny of the expert, who thus will become a decision-maker without being answerable to the community. Participation of the public in the decision-making process involving scientific problems is probably essential to the coherence of democratic societies and to the survival of their institutions.

A parenthesis might be opened here for a brief discussion of the practical problems posed by the teaching of science to nonscience students (or to the layman). H. R. Crane, a professor of physics at the University of Michigan, has this to say in his article, "Students Do Not Think Physics is 'Relevant' " (1968):

> The noncalculus students come to us with a pretty highly developed language, and they even have their own terms for many phenomena of physics. The trouble is, it is different from ours. It would be fine if we could start at the beginning of the course discussing physics in their language and gradually convert to ours by the end of the course. Instead, we feel that we have to start right off using ours. We are like Parisian taxi drivers. We don't recognize any tongue but our own. Their only defense is

to repeat the words and phrases back to us, but when they do that we are not having two way communication.

Another difficulty comes from teachers of science who commonly take the position that the most or indeed the only important aspects of knowledge are those which are now being developed and preferably can be formulated in theoretical language. Here again it seems useful to quote from a speech by the Columbia University physicist Polykarp Kusch, Nobel Prize laureate:

Man's principal contact with nature is through the immediate, visible world, as seen by primitive man and as modified by contemporary man. There is nothing intellectually shabby in the attempt to describe and explain the immediately observable world. How do the devices and procedures of a technological society work? This kind of question has a bad reputation; it is described in the pejorative phrase, household physics. I think the student would find both value and intellectual excitement in attacking the problem of why you need press only one button to heat your apartment and another to cool it. How does it all happen, starting with the peculiarly sulfurous fuel that Consolidated Edison burns in New York? None of this is a part of the current excitement of science, but it has an enormous content of scientific knowledge and deals with the visible realities of the world of science.

Professor Kusch's remarks are related to the fact, noted in an earlier chapter (pages 24, 36), that many scientists are more interested in the *advancement* of knowledge, than in its possession.

The layman's acquaintance with scientific knowledge need not be limited, of course, to those aspects of science which have practical applications and relevance to social problems. The kind of knowledge which is of greatest interest to the layman, however, differs in nature from that pursued by the specialists. Outside their professional activities, most persons, scientists included, are more concerned with purposes than with ultimate causes or detailed mechanisms. In the presentation of theoretical science to the general public, the explanation of particular aspects of phenomena must focus on the meaning of the whole.

The belief that almost any kind of broad knowledge can be communicated to intelligent persons has led the French Academy of Letters to admit into membership many scientists and technologists known for achievements in areas foreign to literature. As the philosopher Ernest Renan wisely stated in his speech welcoming Louis Pasteur at the Académie Française, "Everything becomes literature when it is done with talent." Our societies would not long survive if the multifarious knowledge of specialists could not be integrated in the general thought processes of society.

The language problems posed by the communication of scientific knowledge to the lay public were already being discussed in England during the seventeenth century. The first association of scientists in England, The Royal Society, was founded in 1662 during a period when the English language was overloaded with rhetorical redundance and unnecessary fantastic images. The early members of The Royal Society were so conscious of this "superfluity of talking" that they committed themselves to reporting their discoveries and thoughts in a "close, naked, natural way," and with "clear senses." Their taste with regard to styles of speaking and writing has been recorded by Thomas Sprat in the first *History of the Royal Society* that he published in 1667. The Fellows of the Society, according to Sprat, resolved "to reject all the amplifications, digressions and swellings of style," and to prefer "the language of Artizans, Countrymen, and Merchants, before that of Wits or Scholars."

The simplification of the scientific language occurred rapidly, not only in England but all over Europe. It enabled scientists to communicate their specialized knowledge—often with great success—to the general public. At the end of the seventeenth century and throughout the eighteenth century, people came to Paris from all over Europe to attend public lectures and demonstrations on various aspects of science. In Germany the illustrious physicist H. von Helmholtz found it possible, in a number of essays that are still worth reading, to express complex problems of theoretical science in a language meaningful for the layman. During the nineteenth century, the meetings at the Royal Institution became a fashionable London rendezvous.

The Christmas Lectures for a juvenile audience delivered at the Royal Institution by its director Michael Faraday and his followers admirably illustrate that precision of thought, clarity of language, and felicity of expression can go far toward describing and explaining scientific facts and theories without the use of professional jargon. The popularization of science both by scientists and science writers is still a flourishing enterprise in our own times. Many books on science for the general public published during recent decades have become best sellers, even though they deal with theoretical themes of anthropology, biology, physics, astronomy—even of mathematics—unrelated to the practical affairs of ordinary life.

The simple language that had been advocated by the founders of The Royal Society is still adequate for the communication of much scientific knowledge. The vocabulary of science has grown of course, but this growth has been chiefly through the steady expansion of the range of household words which derive their meaning from the direct apprehension of reality through the senses. Certain aspects of modern science, however, have brought us face to face with the realization that ordinary language is inadequate to deal with concepts having no reference to perception by the senses. For example, phenomena occurring at the subatomic or cosmic levels can be described to the layman and the non-specialist as well only by using far-fetched analogies—a technique of description that inevitably entails great limitations and dangers of misunderstanding. Such words as waves, particles, conservation, symmetry, and freedom are used by Physicists to denote concepts having little if any relevance to the meaning that the same words have in common usage. The concepts of theoretical physics are abstract formulations of the intellect that do not correspond to anything perceived by the senses; there is no way to convey their meaning to non-specialists except through the suggestive value of approximate images.

Lord Ritchie Calder, one of the most skilled exponents of science for the layman, has recently described both the difficulties and possibilities of his trade (1964):

"You say you want an explanation of Einstein's Theory of Relativity. What kind of explanation? In terms of words of the Anglo-Saxon period and therefore with very nearly the concepts prevalent at that time? In terms of the language of the seventeenth century and therefore with concepts prevalent about the time of Newton? In terms of the language of, say, 1900? In modern technical terms? In modern mathematical symbolism? All these would represent attempts at explanation but how successful could they possibly be?" Or, this writer might add, in terms of a London bar where, over a glass of beer and bread and cheese, Professor Levy himself once gave me a "translation" of Einstein, which made me sound very convincing that night on the BBC. I admit it was rather like getting a Cree Indian to define atomic energy–which I once did. I found government geologists teaching Red Indian trappers in the Canadian North to look for uranium ore, and I wondered what it all meant to the Cree Indians. I asked the chief of the Crees, what, in his language, was atomic energy. He replied: "Eskotik-otchit kaochipyik," which means "Lightning which comes out of rock."

Ordinary language is a household invention which is not applicable far beyond the confines of household experience. By the skillful use of analogies a few writers have nevertheless succeeded in conveying some awareness of abstract theoretical concepts, but there is danger in applying this awareness to problems in other fields. This is illustrated when the immensely sophisticated physical and mathematical concepts of relativity and uncertainty are invoked—because the words have a familiar ring—to defend trivial concepts in philosophy and sociology.

A particular science is most readily understandable to the general public in the early phases of its development when it is still dealing with phenomena not too remote from ordinary experience. This is probably the reason why Darwin's hypothesis made such a rapid and profound impact on modern thought. It is also possible, as pointed out by J. Robert Oppenheimer (1962), that "scientific knowledge will resonate and change the thinking of men only when it feeds some hope, some need that pre-exists in society." Still in Oppenheimer's words:

> The hunger of the Eighteenth Century to believe in the power of reason, to wish to throw off authority, to wish to secularize, to take an optimistic view of man's condition, seized on Newton and his discoveries as an illustration of something which was already deeply believed in quite apart from the law of gravity and the laws of motion. The hunger with which the Nineteenth Century seized on Darwin had very much to do with the increasing awareness of history and change, with the great desire to naturalise man, to put him into the world of nature, which pre-existed long before Darwin and which made him welcome. I have seen an example in this century where the great Danish physicist Niels Bohr found in

the quantum theory when it was developed thirty years ago this remarkable trait: it is consistent with describing an atomic system, only much less completely than we can describe large-scale objects. We have a certain choice as to which traits of the atomic system we wish to study and measure and which to let go; but we have not the option of doing them all. This situation, which we all recognise, sustained in Bohr his long-held view of the human condition: that there are mutually exclusive ways of using our words, our minds, our souls, any one of which is open to us, but which cannot be combined: ways as different, for instance, as preparing to act and entering into an introspective search for the reasons for action. This discovery has not, I think, penetrated into general cultural life. I wish it had; it is a good example of something that would be relevant, if only it could be understood.

While the popularization of scientific facts and laws is a highly worthwhile enterprise, it may not constitute the best prescription for improving public understanding of science. To this end, better appreciation of the processes through which science develops may be more effective than disseminating scientific information. As stated by the American chemist and educator, James B. Conant, former President of Harvard University: "Being well informed about science is not the same thing as understanding science, although the two propositions are not antithetical. What is needed is methods for imparting some knowledge of the tactics and strategy of science to those who are not scientists."

Unfortunately, it is not possible to give a universally acceptable formulation of the scientific method, because opinions on this score differ greatly among scientists themselves. Some of the most eminent among them go as far as stating that there is no such method. Nor is there any agreement as to what components of the scientific enterprise should be introduced in the education of non-scientists.

Modern man, it is often asserted, can no more afford to be ignorant of science than medieval man could ignore the Christian Church or the feudal system. But while the wording of this analogy is picturesque, its meaning is far from clear. Medieval man did not read Thomas Aquinas or the Church fathers, nor did he know feudalism as an economic or political system. For him Christianity was the threat of the tortures of hell or the promised joys of heaven; it was the mystic emotion created by stained glass windows, identification with legend through church statuary, and the participation in religious ceremonies. Feudalism meant turning part of his meager income over to the lord in the manor and fighting his wars.

With few exceptions the modern layman knows merely a few artifacts and by-products of science. He does not appear to be either more able or more inclined to concern himself with scientific abstractions than medieval man was able to think about theological problems or about the sociology of the feudal system.

There is no criterion, furthermore, to guide selection as to what kind of scientific facts would be suitable for the education of the layman. When Sir Charles Snow first discussed *The Two Cultures* in his celebrated Rede Lecture of 1959, he stated that an understanding of entropy and of The Second Law of Thermodynamics was the common property of all educated men and an essential part of modern culture. But when he reprinted the same lectures a decade later, he emphasized instead knowledge of molecular biology, the DNA molecule, and its relation to heredity. It is not unlikely that he might still change the priority if he were republishing his lecture today. For example, he might judge that the urgency of such problems as overpopulation, food shortages, violence in human relationships, and deterioration of the environment makes it imperative for the citizen to have some understanding of demography, ecological principles, or general systems theory. Molecular biology is still the most fashionable field of biology, but such problems as health, socialization, or behavior elude the molecular approach, because they deal with phenomena that occur at higher levels of complexity. The dynamic concepts of organization and function are much more *à la page* than those involving energy and structure.

Rapid and profound shifts of emphasis have repeatedly occurred in the scientific community, in part because fashions change in science even more than in other types of endeavors, also because social concerns inevitably affect intellectual preoccupations. J. Robert Oppenheimer had such shifts of scientific interests in mind when he wrote the passage quoted above.

Knowledge of scientific facts, even the important ones, is not sufficient to provide understanding of the tactics and strategy of science. This may require direct personal involvement in some aspect of the scientific enterprise. Even more difficult perhaps is an appreciation of the complex interrelationships between theory and practice.

In his classic book *Science and Commonsense* (1951), James B. Conant recognizes similarities between practical inventions and theoretical discovery, but he also points to important differences between these two forms of scientific research. Other students of the sociology of science have introduced further refinements in this distinction. The adjectives "theoretical," "basic," or "fundamental" are used to designate these aspects of science that are concerned with knowledge per se, regardless of its relevance to practical applications. "Mission-oriented" science refers to a kind of research which has a societal rather than technical goal. "Applied" research or development is focused sharply on well-defined technical goals.

The expressions used to qualify research are useful in differentiating the various levels of abstraction and practicality at which the scientific enterprise is conducted. For this reason they are now common currency, especially in the executive and legislative branches of government responsible for the organization, administration, and financing of scientific research. In practice, however, it is not always possible to separate clearly the basic from the applied aspects of science. Scientific studies that could be considered as very theoretical in a school of medicine, pharmacy, agriculture, engineering or architecture might well appear as mission-oriented or even narrowly applied among academic scientists working on abstract questions at the frontiers of understanding in the laboratories of an ivory tower. The studies required for sending man to the moon are called scientific by NASA, but not by cosmologists concerned with the origin of the universe. Similarly, biological research on new methods of contraception is considered theoretical science by demographers but not by biologists whose primary interest is to determine how inanimate matter first acquired the attributes of life.

Everyone knows that scientific knowledge increases quantitatively with time, but laymen find it difficult to understand that it also changes qualitatively while evolving—to such an extent that concepts assumed to describe reality a generation ago may now appear naive, innocent, or even nonsensical. An important aspect of the layman's science education is to explain that scientific knowledge is never absolute or final, yet it remains valid when considered in the social and intellectual framework within which it was developed.

Scientists of the late seventeenth century developed a comprehensive view of the physical universe, based on new analytical methods and principles but incorporating observable facts and laws established by their predecessors. Their factual knowledge was old, but their interpretations were new. Yet, these interpretations now appear inadequate to us. Contemporary scientists have absorbed seventeenth century physics and transformed it into new revolutionary concepts, which in turn will certainly be used in the future for more sophisticated concepts.

Laymen are disturbed by the frequent expressions of disagreement among equally competent scientists on practical problems involving scientific knowledge. This naturally makes them question the objectivity of science. It is important, therefore, to make clear that scientists hardly ever disagree on the validity of the facts themselves, but only on the interpretation and use of these facts.

In some cases disagreement among specialists has to do with the amount of factual evidence required to justify the acceptance of a hypothesis. For example, the relation of cholesterol blood levels to vascular disease is supported by a few facts, but the evidence is not sufficiently convincing to be accepted by all physicians. In other cases disagreement concerns the application of well-established scientific facts to social problems. No one doubts that ionizing radiations increase mutation rates, that most mutations are deleterious, and that some damage to human life is therefore likely to result from an increase in radiation level. But, even though all scientists agree on these facts, individually they differ as to the levels of radiation they consider to be tolerable, because this involves social considerations unrelated to the theoretical problems of radiation and health. For example, the testing of nuclear weapons or the industrial use of nuclear energy will unquestionably raise somewhat the radiation levels in the environment, but the biological hazards of this change may be considered socially less important than the increase in the nation's military and economic strength.

The current discussions concerning the advisability of banning the use of DDT and other chlorinated pesticides illustrates how complex it is to evaluate even the usefulness of a simple substance. DDT is an immensely effective insecticide that causes little if any damage to man when used under controlled conditions. Its toxicity becomes evident chiefly after prolonged periods of massive usage when it has progressively accumulated in the food chains.

All technological innovations similarly have delayed and indirect consequences, which are almost independent of their initial effects. But human life can be profoundly affected—for good and evil—by such indirect delayed consequences. Innovations must therefore be judged not only from the point of view of their effect in the here and now, but also with regard to the possibility that they may damage human life or the environment, or both, at some later date. Scientists themselves frequently overlook this possibility. Educating the general public to the fact that all aspects of life and the environment are interrelated and that every intervention is likely to have unforeseen consequences might have the useful result of compelling the scientific community to focus its attention on complex ecosystems.

The need for public awareness of the role that science could (but does not) play in a better management of the modern world is highlighted by current discussions of the environmental crisis.

The Federal Government is committed to a program of social and environmental improvement; the American Institute of Planners has conducted several extensive symposia focused on *The Optimum Environment* (Ewald, 1967); the Ford and Rockefeller Foundations have announced their intention to devote a large percentage of their resources to problems of human ecology. The intent of all these enterprises is essentially to develop social and technological practices to solve the problems of man in the modern world. The common limitation of all these well-meaning projects is the lack of knowledge concerning the long-range effects of environmental factors on the well-being and development of man. For example:

(a) Everyone agrees that it is desirable to control environmental pollution. But what are the pollutants of air, water, or food that are really significant? Sulfur dioxide, nitrogen oxides, carbon monoxide, and the other volatile products of automobile exhausts are commonly regarded as the major components of air pollution. But what about the immense amount of colloidal particles that also contaminate the air of our cities, for example, the asbestos particles released from brake linings and by the construction trade or the material released by automobile tires grinding on street surfaces? The acute effects of pollutants can be readily recognized, but what about the cumulative, delayed, and indirect effects? Does the young organism respond as does the adult? Does he develop forms of tolerance or hypersusceptibility that affect his subsequent responses to the same and other pollutants? Priorities with regard to the control of environmental pollution cannot be established rationally until such knowledge is available.

(b) Everyone agrees that our cities must be renovated or even rebuilt. But while technologies are available for almost any kind of scheme imagined by city planners, architects, and sociologists, no one knows how the environments so created will affect health and behavior and especially how they will influence the physical and mental development of children. We know how to create sanitary environments that permit the body to become large and vigorous. But what about the effect of environmental factors on emotions and on the development of the mind? All too often housing developments are designed as if they were disposable cubicles for dispensable people.

(c) Everyone agrees that all citizens should be given the same educational opportunities. But what are the critical ages for the development of mental attributes and for receptivity to the various kinds of stimuli? To what extent can the effects of early deprivations be corrected? We need a scientific knowledge of the effects that prenatal and early postnatal influences exert on the biological and mental characteristics of the adult.

These three classes of examples illustrate that environmental control has been considered so far almost exclusively from the point of view of technology, in ignorance of the responses that the organism makes to environmental forces and without regard to distant consequences for human welfare. Whether approached from the scientific or practical point of view, environmental improvement must take into consideration both the constraints and the formative effects of the environment, not only in the present but also in the future.

The formulation of these problems requires social judgment as much as specialized technical knowledge. Perceptive and informed laymen can therefore contribute to such formulation at least as much as professional scientists. Indeed, laymen may have an advantage over scientists because their overall view of human problems is not distorted by the parochialism which commonly results from technical specialization. As Edmund Burke wrote in "Reflections on the Revolution in France": "When men are too much confined to professional and faculty habits, and as it were inveterate in the recurrent employment of that narrow circle, they are rather disabled than qualified for whatever depends on the knowledge of mankind, on experience in mixed affairs, on a comprehensive, connected view of the various, complicated, external, and internal interests, which go to the formation of that multifarious thing called a state."

The fact that the social performance of scientists is increasingly coming under public scrutiny will inevitably affect public education. The technical and theoretical achievements of science should be communicated to the lay public as accurately as possible, yet it is perhaps more important that the general and practical implications of science be defined and discussed. The philosophical and social quandaries raised by scientific progress must be publicly emphasized just as much as the prospects of technological breakthroughs. New kinds of economic, educational, and ethical problems are daily being created by scientific innovations. Society can assess them only if it possesses some awareness of their general significance and potential consequences. The popularization of science, whether by practicing scientists or professional writers, must therefore transcend the publicity given to the latest knowledge of technical achievements. It must be concerned more with the social, political, and even moral aspects of science policy than with a "gee-whiz" enthusiasm for spectacular or startling feats.

Understanding the relationships among science, technology, and social problems is not the exclusive prerogative of professional scientists; it can be acquired by educated citizens. The public debates on the social and economic aspects of science now going on in and out of Congress will certainly compel a sharper definition of the comparable roles of the expert and of the citizen in the formulation of scientific policies.

Increasingly, a large percentage of our population will receive college training and will, as a result, demand scientific expertise in socio-political decisions. But it would be disastrous if respect for expertise resulted in an increase in docility before experts. Freedom can be maintained only if citizens understand the intellectual basis of scientific expertise sufficiently well to differentiate between persuasion and manipulation by experts.

All important human activities have given rise to a highly sophisticated profession concerned with the criticism of their values, achievements, trends, and potentialities. The word criticism now commonly denotes an unfavorable judgment, but it is used here in the more creative sense of its Greek etymology; *kritikos* denotes the ability to discern or judge. In our societies the professional critics of art, music, literature, economics, and government play a creative role even when they do not themselves contribute directly to the fields of activity that they evaluate. Some of the most eminent critics of art or of government have never composed anything or held public office. Science would benefit from the kind of evaluation that professional critics give to other human activities. This is probably what the Abbé Saint Pierre had in mind when he proposed at the end of the eighteenth century the establishment of a body of "scientific politicians for planning purposes." (Quoted in Butterfield 1950)

Whether scientific criticism should develop from within the community of experimental scientists or outside it remains a moot question. Upton Sinclair's *The Jungle* was a novel, yet it was most effective in influencing Congress to pass the Food and Drug Administration Act. *Silent Spring* by Rachel Carson was not written by an entomologist, nor was *Unsafe at Any Speed* (Ralph Nader) by a Detroit automobile engineer. For reasons that are still unclear, persons trained in the orthodox scientific disciplines rarely contribute to the formulation of the environmental problems that are of most concern to citizens. They find it difficult to be part of the scientific establishment without becoming its prisoner. It is certain, in any case, that the higher criticism of science cannot have much vitality without public participation.

A society that blindly accepts the decisions of experts is a sick society on its way to death. The time has come when we must produce, alongside specialists, another class of scholars and citizens who have broad familiarity with the facts, methods, and objectives of science and thus are capable of making judgments about scientific policies. Persons who work at the interface of science and society have become essential simply because almost everything that happens in society is influenced by science.

6
The Willed Future

□□ Lewis Carroll was professor of mathematics at Oxford University when he wrote in *Alice's Adventures in Wonderland*:

"Cheshire-Puss. . . . would you tell me please, which way I ought to go from here?"

"That depends a good deal on where you want to get to," said the Cat.

"I don't much care where . . . " said Alice.

"Then it doesn't matter which way you go," said the Cat.

" . . . so long as I get *somewhere*," Alice added as an explanation.

"Oh, you're sure to do that," said the Cat, "if you only walk long enough."

The Cat's answer has often been quoted to express the view that scientists do not know where knowledge is taking mankind, and furthermore they do not really care. Science, it is said, cannot provide social goals because its values are intellectual, not ethical; once social goals have been selected by non-scientific criteria, science can determine the best way to proceed. Yet it is probable that science can help in formulating values, and thus in setting goals, by making man more aware of the consequences of his actions. The need for knowledge of consequences in decision-making is implicit in the Cat's remark that Alice would surely get *somewhere,* if she would only walk long enough. Since this *somewhere* might turn out to be a very undesirable place, it is best to make conscious choices as to where one wants to go.

In Lewis Carroll's time, at the end of the past century, most scientists were like Alice, unconcerned as to where they ought to be going. They saw no point in formulating social goals for their professional work, because they regarded science as an end rather than a means. Some of them took an even more cavalier view of their relation to society and boasted of the fact that their work could never be—as they erroneously thought—of any practical use (Hardy, 1940). Hardy's toast mentioned earlier, "Here is to pure mathematics, may it never have any use" symbolized a widespread desire to keep science as a high class diversion to be pursued and valued for its own sake. The truth is, of course, that one of the characteristics of science is its incapacity to be impractical. The scientists who realized that nineteenth century science was beginning to transform human life, did not worry about the social implications of this fact, because they took it for granted that practically all scientific advances would eventually yield some useful technique or product. In consequence, they regarded scientific and technological feasibility as the sole determinant of decision and action. Unconsciously and insidiously the word "can" thus came to imply and replace the word "ought" in the vocabulary of scientists, technologists, and even of sociologists.

Since the seventeenth century, science has been identified with economic and social progress. Whether Francis Bacon really played the most important role in giving a new direction to the scientific enterprise—as his admirers believe—or whether he was merely the eloquent spokesman for opinions which were prevalent around him is a debatable question. It is certain, however, that he made it intellectually and socially fashionable for scientists to take an active part in the practical problems of their times. The members of the "invisible college" who joined in 1662 to form the first great English scientific institution, The Royal Society, had no doubt as to what science was for. It was for practical use. Robert Boyle wrote to one of his friends in 1646, "Our new philosophical college . . . values no knowledge, but as it hath a tendency to use." The Royal Society was spoken of in the dedication to Charles II as "a perpetual succession of inventors." In fact, during the seventeenth and early eighteenth centuries the technological requirements of English economy greatly influenced the choice of the scientific problems on which the "virtuosi" decided to concentrate.

English science later became more aristocratic and theoretical. It emphasized the discovery of the facts and laws of nature, regardless of relevance to human affairs. By the end of the eighteenth century The Royal Society was so remote from the social applications of science that this field was left open for the practical genius of Count Rumford, an American who had been employed by the King of Bavaria. Rumford founded in London The Royal Institution "for bringing forward into general use new inventions, and improvements, particularly such as related to the management of heat and the saving of fuel and to various other mechanical contrivances, by which domestic comfort and economy can be promoted."

Like The Royal Society, The Royal Institution soon became involved in theoretical science. For example, it provided the laboratories in which Humphrey Davy and Michael Faraday made their fundamental discoveries. By 1831 it had become a meeting place of the intelligentsia and was less and less committed to the people. Eventually, this change of emphasis led to the establishment of the British Association for the Advancement of Science, which was expected to place greater emphasis on the social applications of knowledge. During the past two decades this organization has pioneered in the debates concerning the social implications of science and the social responsibilities of scientists—an example which has been followed by other scientific associations all over the world.

The social history of science in the United States is not entirely parallel to that of science in England because of social and economic differences between the two countries. The first institution for scientific research in the American colonies was probably the American Philosophical Society, organized in Philadelphia by Benjamin Franklin in 1743, "for the promotion of useful knowledge." Throughout the nineteenth century, science in America retained this practical bias. It was "sold" to the public on the basis of its contributions to important American values—utilitarian, equalitarian, religious, and even as a means of social control. (Daniels, 1967) In the age of Jackson, with the powerful pressures on the citizen to do useful work, scientists had to greatly exaggerate the practical utility of their work, because few persons believed in it at that time. The free inquiry into nature divorced from utilitarian purposes achieved some social recognition only after the Civil War.

By the late nineteenth century the technological applications of theoretical science became evident to everyone, especially those in the chemical and electrical industries. These achievements secured support for science in general, but the emphasis remained on its practical aspects. Attacks on pure research as non-productive were common in the late nineteenth century, and a congressional committee (the Allison Commission) was established to study the administration and organization of scientific agencies. There were profound cuts in the appropriations for scientific bureaus in the 1880's, and several research-oriented scientists were dismissed.

The two world wars, and especially the Sputnik

era, brought about a widespread recognition of the need for theoretical research and resulted in the fantastic expansion of the scientific enterprise we have just witnessed. (Bush, 1945) There are indications, however, that the pendulum is swinging again and that "useful" knowledge is once more the order of the day.

The ambiguity of social attitudes toward science may account in part for some disconcerting behavioral traits of the scientific community. Scientists are prone to teach their students that science is a great intellectual adventure to be pursued without regard for practical utility. But they encourage technologists and popularizers to convey a different message to the public because they realize that science for science's sake does not have enough general appeal to secure large financial support. The very same scientists who affirm in academic circles that they are primarily concerned with eternal truths publicize the practical potentialities of their work when they testify before congressional committees. This conflict between the pure science ideal and the exigencies of democratic processes has profound and unhealthy consequences. In particular, it makes the planning of scientific research much more responsive to publicity than to scientific logic or to real social needs. Even the would-be "pure scientists" are increasingly influenced by non-intellectual criteria in the selection of their research problems.

Many university scientists would soon abandon their present lines of work if they were to lose the financial support they receive as a result of socially conditioned fears and compulsions. This is obviously true for the huge and costly programs in atomic physics which were started during the war, not for the sake of knowledge, but rather to surpass the Germans. These are now continued chiefly for reasons of international politics. The space program is likewise nurtured less by intellectual curiosity than by the desire to surpass the Russians. Financial support for molecular biology has also been largely obtained on a questionable scientific ground—namely, the unwarranted affirmation that the control of cancers, vascular diseases, and mental disorders will come from detailed chemical knowledge of the structures that make up the body and the brain. And so it goes throughout the whole scientific enterprise.

Individual scientists exhibit intellectual integrity in their professional work, but as a group we tend to abandon intellectual and ethical discipline when unwarranted claims give promise of increasing social support for the problems we happen to find interesting. Dedication to the discovery of truth for truth's sake seems to be quite compatible with the more mundane desire to work in fashionable fields, preferably those which are well financed and likely to be rewarded by academic promotion and glamorous prizes—let alone by plump consultant fees.

The manned space program is the most ambitious and most sophisticated technological enterprise of our times, yet there was little scientific rationality in its promotion. President Kennedy was naturally motivated by reasons of international prestige when he decided that a huge percentage of the national effort should be dedicated to landing a man from the United States on the moon by 1970. But the eminent scientists who supported the project defended it with less valid intellectual reasons. When asked to explain why it was important to land a man on the moon, they could only quote George Mallory's answer when he was asked why he had been so eager to climb Mount Everest: "Because it is there."

As an explanation, the answer that the moon must be explored because it is "there" is a perfect panchreston (an explain-all) that could be used for any decision, however trivial. There are many things in the world that are at some "there" and that would present the human spirit with challenges at least as great as those presented by lunar exploration. The ultimate constituents of matter, the vagaries of the weather, the origins of man, the way he learns to speak, starvation throughout the world, and the soul-destroying ugliness of our cities are also problems which are very much "there," crying out for study and for action.

The scientific establishment is shockingly irrational in the selection of its priorities and in the determination of the relative amount of support it gives to various fields. Certain problems are emphasized because their supporters have political influence or can appeal to popular emotions. Other problems in contrast are neglected even though their study would enlarge understanding of the cosmos or contribute to human welfare. Two examples, one taken from the biomedical sciences and the other from engineering, will illustrate the kind of discussion that should go on within the scientific community to make the selection of its programs of research and teaching more rational and better adapted to human needs.

Everyone knows that surroundings and ways of life profoundly influence the biological and mental development of children and thereby condition the type of adults they will become. Even if we succeed in providing all children with peace and economic affluence, they will certainly suffer as adults from the technological, environmental, and social absurdities to which they have been exposed during their early life. We know little if anything of the effects exerted by such early influences beyond the fact that these effects are profound and lasting.

Medical schools and research institutes are primarily concerned with the treatment of disease, rather than with the knowledge of normal development. Furthermore, present-day biological research tends to focus its efforts on the study of simple biological systems, preferably those that can be reduced to the cellular or subcellular level. Yet, the most important biological and mental characteristics of man emerge from phenomena that cannot be studied by this approach, since they result from the responses made by children to the environmental and social conditions that we select or create for them. The interplay between the developing organism and his total environment and the long-range consequences of this interplay constitute, therefore, some of the most urgent problems of human societies. But knowledge concerning them will not develop as it should until we change the scientific philosophy of existing biomedical institutions or preferably organize new ones dedicated to research and teaching focused on a holistic science of man.

For lack of personal experience, I would hesitate to express an opinion about the need to change the emphasis of scholarship in such fields as architecture or engineering. Fortunately, this need has been clearly stated by a graduate student in engineering in a recent article entitled "Students, Technology, and New Priorities" (1968). I shall therefore let John Wirt, the engineer in question, speak for me as well as for many of his contemporaries.

In the School of Engineering, internal debate arises from the conflict between the recognized needs of our society and the traditional output of technology. One result is an inclination within the School to direct technological endeavor away from its historic preoccupations, and toward the solution of pressing social and managerial problems. Professors are shifting the subject of their research, students are lobbying for new educational experiences, and new programs and departments are emerging.

The social forces causing this conflict have, almost ironically, been created by the immense material successes of technology. Only a small portion of the productive capacity and reserve of capital in this country is needed to satisfy each citizen's basic shelter, food, and health needs—leaving extensive resources available to create comfort goods. While limited resources prevent the attainment of every desirable goal, the range of possible alternatives is still great. The challenge to educators and students is to help society make the right choice.

The social forces which have created debate in the university's technological community are both external and internal. From the outside, federal and state agencies are beginning to redirect research and development funds to social systems analysis, urban research, and social development programs. Budgetary pressures are forcing researchers into these new areas. The government is also compelling technologists to consider the external consequences of their work, and technologists are responding with new developments in methods of analysis. For example—superhighways through cities enable a rapid flow of goods and people; however, they breed air pollution, noise, and unsightly conditions for neighboring non-users. Technologists are now being challenged to construct highways *and* cope with any negative effects.

These challenges are grasped and reflected internally by the heightened social conscience of a number of engineering students and faculty members, and have led them to question the social worth of certain educational programs.

The question we all ask is whether technology will have any success in its ventures into social problem solving. The positive-thinkers believe that the deductive, analytical and managerial methods of the technologist can and must be employed if our increasingly complex systems are to be efficiently managed. They believe, for example, that the quality of life in urban areas can be improved by an efficient and esthetic layout of city functions—from education and recreation to transportation and waste disposal—with direction from professional problem solvers.

To be effective in this broadened sphere of influence,

the technologist must acquire new skills. Foremost is the ability to cope with and incorporate *human and subjective values* into the design process. In the past, engineers have generally eschewed *esthetic factors and social judgments* in their constructions. In the future, technical analysis must be responsive to *societal goals* and values, not only to the engineer's *parochial satisfactions.* [italics mine]

The use of phrases such as "subjective values," "esthetic factors," and "social judgments," symbolizes a shift away from exclusive concern with "the engineer's parochial satisfactions." The emphasis on the role of governmental and other social forces in the evolution of engineering activities suggests that many fields of science will increasingly be influenced by parascientific factors. As pointed out by Alvin Weinberg (1968): "It is a venerable philosophic principle that the value of any universe of discourse must be judged from without that universe of discourse. It was for this reason that . . . I urged that large-scale public support be given a field of science *only if* it rated well with respect to what I called 'external criteria.' "

Scientists find it intellectually unjustified and dangerous to introduce subjective values in their professional activities. One reason for this attitude is that values usually involve complex situations not readily amenable to scientific analysis, such as relationships among human beings, religious or esthetic experiences, and judgments as to what is desirable or not. Furthermore, values imply freedom and therefore cannot be entirely accounted for by scientific determinism.

There may exist some values which are absolute

either because they are determined by the universal order of things or have been woven into the fabric of man's nature during the course of his evolution. In practice, however, most values by which men operate are derived from prevailing social attitudes, beliefs acquired early in life, and the various experiences which are peculiar to each person. Other values probably originate from the awareness that certain courses of action have desirable or undesirable consequences—an awareness acquired in the course of evolutionary and personal development. There is a real possibility that values can be modified by the natural and social sciences in the future, provided the scientific community is willing to concern itself with the indirect and distant consequences of its activities. Science, Thomas Hobbes wrote in *Leviathan* (1651), is the knowledge of consequences. In most cases, admittedly, science is still far from providing an exact knowledge of consequences, but it can greatly facilitate their anticipation.

Scientific knowledge per se is not sufficient to formulate the values that govern human behavior, nor can it impose them on society. However, it can provide a more factual basis for options by giving the statistical probability that certain consequences will result from new technological and social practices. Since awareness of likely consequences plays a large role in decision-making, scientific knowledge can become an important criterion in the evaluation of old value systems and perhaps in fostering the development of new ones.

In a television broadcast before the BBC, Ritchie Calder provided a striking example of the need—especially for statesmen—to consider the ultimate consequences of scientific technology in important social decisions:

The mass-audience of British television heard the heart-cry of Clement Attlee, the Prime Minister who had concurred with President Truman in the decision to drop the bombs on Japan. "All I knew was that it was a bigger bomb. I knew nothing at all about fall-out, nor the genetic effects. And as far as I know President Truman and Winston Churchill knew nothing of those things either. Whether the scientists directly concerned knew, or guessed, I do not know. But, as far as I am aware, they said nothing of it to those who had to make the decision. I am no scientist you know." (R. Calder, 1964)

And yet, H. J. Muller had been awarded the Nobel Prize eighteen years before Hiroshima, in 1927, for his discovery that radiations cause genetic mutations.

The lesson of Hiroshima has not been entirely lost. Immediately after the end of the Second World War, and therefore long before practical techniques had been developed for the industrial use of nuclear energy, the National Academy of Sciences in the United States, and several other scientific organizations throughout the world, began to study the biological effects of ionizing radiations. The knowledge thus acquired has profoundly influenced international policies and the civilian uses of nuclear energy, because it made scientists and the general public aware of the grave biological consequences likely to result from exposure to radiation. The prospective study of the biological effects of ionizing radiations thus constitutes a precedent for the social doctrine that no new technology should be placed in the public domain until thorough scientific studies have been made of its consequences, in particular of its potential effects on human life and the environment.

The formation of the Food and Drug Administration half a century ago, and more recently the Consumer Protection Administration and other agencies for environmental control, represent attempts at controlling the dangers of new technologies through the prospective study of their immediate and long-range effects. The problem was stated in the report on "Noise and the Sonic Boom in Relation to Man," presented to the Secretary of the Interior in November 1968 by a group of experts. The experts pointed out that "every new technology carries its own adverse effects and these must receive attention equal to the technology itself. Usually the public which reaps the benefits from a decision is different from the public which suffers the adversities. The decision-making should recognize both publics by providing for continuing work on both sides of the question, not at some distant time in the future but right from the moment of decision."

The clumsiness and ineffectiveness of the measures taken so far to control drug usage and to prevent the various forms of environmental pollution should not invalidate the doctrine that knowledge of consequences is an essential part of good technology. To a large extent, the ecologic crisis results from the fact that modern societies have not yet appreciated the need for exhaustive scientific studies of the long-range consequences of social and technological innovations.

Skeptics have good reasons to claim, of course, that knowledge of consequences does not necessarily modify human behavior. They can point out that the publication of the dangers associated with cigarette smoking or high-speed automobile driving has not significantly modified the behavior of intelligent and well-informed people. Such disregard of evidence simply means that most persons are willing to take calculated risks—as everyone constantly does in daily life. Scientific knowledge can help in deciding what risk is acceptable by defining the statistical probability that certain kinds of consequences will occur and thereby making it easier to determine the relative importance of these consequences within the framework of personal values.

On several occasions in the past, in LeRoy's time for example, scientific advances brought about significant changes in the ways of life. But the effects were usually so limited that value judgments had little place in decisions concerning science. In contrast, the changes that are occurring now are profound and widespread. Within the span of a man's life the biological and mental attributes of immense numbers of people in all countries and all social groups are suddenly and simultaneously affected by social and technological innovations that have their origin in new scientific knowledge. The human and environmental problems thus created are so new that they cannot be usefully illuminated by past experience. We worry more about the social effects of science than did Le Roy's contemporaries, because the rate of change is so rapid that widespread and often disastrous consequences become manifest before their causes have been recognized. Concern for consequences re-introduces the need for new criteria in evaluating social and technological innovations.

As already discussed in several chapters, the social and economic problems of a period usually stimulate scientific research relevant to these problems. For example, theoretical studies on the various forms of energy were actively prosecuted during the nineteenth century in response to the needs of the Industrial Revolution; or, during the period when epidemic diseases constituted the most important medical problems, institutes of medical microbiology were established all over the world. Social forces unrelated to the search for knowledge per se influence scientific trends as much today as in the past. Just as the two world wars and the Cold War have been largely responsible for the emphasis on certain aspects of the physical and space sciences, it is probable that the environmental and urban crises will stimulate the ecological and behavioral sciences.

Scientific developments often take the form of adaptive responses to social demands. As in the case of other living organisms the adaptive responses of the social organism tend to correct distortions in the system and to return it to its initial condition. The American physiologist, W. B. Cannon, referred to these responses as homeostatic and considered them to be expressions of "the wisdom of the body," a phrase first coined by the English physiologist Ernest Starling. While adaptive homeostatic responses are commonly useful at the time they occur, they may be deleterious in the long run and may even have fatal consequences. The phenomena of human pathology demonstrate that the wisdom of the body is often a short-sighted wisdom. For example, the production of scar tissues is an adaptive response because it heals wounds and helps in checking the spread of infection. But scar tissue in the liver or in the kidney means cirrhosis or glomerular nephritis; scar tissue may freeze the joints in rheumatoid arthritis or may choke the breathing process in the lung.

Many of the social adaptive mechanisms of a scientific origin similarly can have deleterious effects in the long run. For example:

—Lowering infant mortality may lead to overpopulation;

—Malnutrition or overnutrition during early development elicits metabolic and psychological corrective processes which may result in pathological consequences later in life;

—Elimination of drudgery by labor saving devices may decrease physical exercise below a safe level;

—Complete protection against certain forms of stress may bring about atrophy of natural defense mechanisms; and

—It is even possible that the development of certain kinds of knowledge may contribute to the feeling of irrelevancy or alienation so common in the modern world.

In the animal world under natural conditions, misdirected adaptive responses are rarely of lasting consequence, because natural selection maintains the species in a state of adaptation by eliminating the misfits. Civilized life, however, greatly restricts the effectiveness of natural selection. Human societies must therefore replace the blind mechanisms of natural selection by some form of conscious planning. In fact, one of the characteristics that clearly differentiates man from animals is that he is consciously goal-seeking. Indeed, in most cases he formulates distant goals long before he has developed the means for reaching them and even before he has established their social desirability. He first dreams of flying, then develops techniques that enable him to fly, and finally he cultivates habits that make flying part of his daily life only after airplanes have become readily available. In many respects human life is teleological.

The word "teleology" is commonly used to mean that the future is inherently and inevitably implied in the present state of affairs. In the *Poetics* Aristotle offered a more appealing interpretation of how the future is related to the present when he wrote, "Truth is not what is, but what may be within the laws of necessity and probability." The qualification "may be" is what makes Aristotle's statement so relevant to the human condition. Man can impose a direction on events but only within the limitations imposed by natural laws. Goal formulation is meaningful only to the extent that man can develop feasible means of implementation, but human life is teleological, nevertheless, because man often selects the goals before he has developed the means to reach them.

It is easy to recognize goal-seeking throughout the religious, political, and cultural aspects of human history. The progress of science also is largely an expression of the faith that certain goals should and can be reached. As already mentioned many scientific dreams of mankind have provided a kind of teleological guidance to early scientists and technologists in their search for what could be done within the constraints of natural laws.

The cynic will remark that advertising could be regarded as a teleological activity since it attempts to create a demand even before the need is felt, and it can be effective only if techniques of implementation exist or can be developed. Advertising conditions the quality of life through the kinds of values and activities it promotes, with effects that can be either good or bad. The same applies to the scientific enterprise.

In principle, the function of science in teleological thinking is rather simple. Goals are formulated, the scientific and technological strategies for reaching these goals are explored, a cost-benefit analysis of the various alternatives is made, and finally tactics are developed to implement the strategy that has been selected. In practice, however, unforeseeable difficulties often arise because most innovations have secondary effects which may be irreversible. The one thing that cannot be done with a theoretical discovery or practical invention is to get rid of it. A scientific innovation is like a djin out of a bottle; the one command it will not obey is to "go back into the bottle."

The selection of scientific goals which are socially worthwhile is of crucial importance precisely because we are saddled with discoveries as long as our civilization exists. For this reason it is regrettable that specialists who are usually more concerned with their work than with its consequences increasingly dominate the social scene in scientific civilization. We have slipped into the habit of regarding most social issues as being chiefly technical in character, with the result that experts are allowed, and indeed often expected, to act almost as rulers of our societies. In the words of Harvey Brooks: "Much of the history of social progress in the twentieth century can be described in terms of the transfer of wider and wider areas of public policy from politics to expertise. Often the problems of political choice have become buried in debates among experts over highly technical alternatives."

Fortunately, there are indications that the role of the expert is coming under social scrutiny. Detecting the issues, shaping the alternatives, judging the political receptivity of the community, setting up a hierarchy of goals, and choosing the appropriate political means to attain the goals are all problems which society may become unwilling to trust to the judgment of experts. The Beatles may have expressed our fear of the expert when they sang of the Nowhere Man in the Nowhere Land making abstract plans for Nobody. There is unfortunately much truth in their suggestion that experts and planners are often as blind as can be and see only what they want to see. The more esoteric the knowledge of the specialist, the more likely he is to be concerned with technical feasibility rather than with social significance.

Mistrusting the expert and emphasizing the dangers of technological "fixes" does not imply a hostile attitude toward either science or technology. Instead what is at stake is the maintenance of a social environment in which man's choices and responses can become increasingly conscious of both formulation of goals and means of implementation. It is not sufficient to ask, "Where is technology taking us?" The more constructive attitude is to plan for the kind of science and technology that will help us to get where we want to go.

There is much more to planning than figuring out an effective and economical distribution of available resources and arranging for the efficient conduct of a particular operation. Planning implies a thoughtful formulation of goals, the input of as much relevant information as possible, the creation of a system offering multiple options, and the possibility of reformulating goals as circumstances demand. Planning should allow for continuous feedback between anticipations of possible futures and events as they actually happen. In other words, the sophisticated forms of planning involve a continuously evolving teleological attitude in which ends influence the selection and development of means, the ends themselves having to be reformulated as the program evolves.

Of course any form of planning requires some kind of forecasting, and this also can be done at several levels of sophistication. Forecasting may simply consist in extrapolating from present trends, as is being done extensively in a rather naive manner for future population size, growth of the gross national product, numbers of automobiles, or other gadgets to be produced in a certain lapse of time. Such extrapolations may be stimulating intellectually but they are seldom correct. Forecasting has also taken the form of the so-called Delphi technique, namely attempting to achieve a consensus of expert opinion on the likelihood that certain social events or technological achievements will occur during a certain period of time. Finally, forecasting, may mean setting a goal for a certain time in the future and attempting to provide the means for reaching this goal or an intermediate desired state, a policy widely practiced by governments and large foundations.

For reasons that are not clear, and in any case that are not justified by actual performance, forecasting now enjoys the dignity of an academic profession. Yet, it is obvious that even the most learned and sophisticated groups of forecasters leave out of their calculations factors which play a very large role in shaping human events. For example, the commission organized by the American Academy of Arts and Sciences under the name "Toward the Year 2000" (Bell, 1967) consists almost exclusively of social scientists and a few natural scientists but does not include in its membership any philosopher, clergyman, writer, artist, politician, soldier, architect, engineer, business man, or student. Yet, such people are much more likely than social scientists or natural scientists to generate the forces that will give its shape to the future.

The scenarios of the future presented in such books as *The Year 2000* (Kahn and Wiener, 1968), seem even more unrealistic. They consist of straightforward extrapolations from present scientific and technological trends, as if it were not known that today's practical knowledge hardly ever solves the problems of tomorrow. Improvements in the efficiency of the dynamo did not prepare the way for nuclear energy; the vaccines and drugs which have proved so effective against microbial pestilences do not help against the epidemics of cancer and heart disease, which are now the most important causes of mortality in the countries of Western civilization; the techniques of psychoanalysis or of behavioral conditioning have no place in the control of the mob disorders that threaten even the most civilized societies. The scenarios of the future prepared by modern forecasters rest, furthermore, on two untenable hypotheses. One is that human beings are robots operated by completely logical control mechanisms; the other is that governments pursue rational policies. Fortunately men change their cultural tastes and interests; and tragically, there are always a few Hitlers around the world to give an unexpected shape to social upheavals. Human life is still largely ruled by primitive urges that survive beneath the veneer of civilization.

Under normal conditions, political forces and expediency determine which programs should be started or rejected, postponed, or modified. The role of the practical administrator is to follow in detail the negative or positive consequences of the different courses of action that have been chosen and to correct errors or unforeseen accidents. Essential as it is, this traditional approach to administration is limited in its possibilities because it is usually restricted to considerations of the "extended present," rather than dealing with the problem as a whole and incorporating visions of possible futures.

Any social problem involves not only individual persons but even more their interplay and their relationships with the total environment, not only in the present, but also in the past and the future. History, environmental forces, value systems, and aspirations all play some role in the practical management of human affairs. In fact, these very factors constitute the determinants of culture for any particular social group. In James Baldwin's words (1963), "A tradition expresses, after all, nothing more than the long and painful experience of a people; it comes out of the battle waged to maintain their integrity or, to put it more simply, out of their struggle to survive."

Because of their complexity, the scientific and cultural aspects of social problems do not fit well in the activities of practical planners and administrators, except to the extent that they shape their attitudes and therefore affect their decisions. In theory, academic institutions would seem to provide a congenial home for the study of social problems free of the limitations and distortions inevitably imposed by practical life. But this is not the case under present conditions.

Universities and colleges are organized to teach and study subjects bearing on almost every conceivable aspect of knowledge. However, they rarely provide the opportunity for integrating the separated fragments of information and for relating them to the needs, values, and aspirations that play such a large role in human life. Yet, it is obvious that all present human problems —racial conflicts, economic growth, the ecologic crisis, the delivery of medical services, the urban blight, environmental pollution, housing and transportation, noise abatement, oceanographic research, and so forth —cannot be understood and dealt with effectively without appealing to multiple specialties of natural, behavioral, and political sciences. The integration of knowledge cannot readily be achieved within the university structure, which is discipline–oriented and not mission–oriented. Yet, integration of knowledge is essential for achieving a comprehensive view of the world in which we live, and it is even more important for the world we want to create.

It would be of immense educational value for students to learn to examine issues with the help of information derived from many different areas of knowledge, but bearing on a particular social topic. The development of skill in assembling and integrating information dispersed through many different academic disciplines would make apparent the interrelatedness of things and the consequent complexity of all social problems. Judgment in formulating social goals must embody both the possible and the desirable and take into account the exciting diversity of the human condition.

The study of science in human affairs inevitably implies a constant emphasis on the interplay between the natural and social sciences. However, this does not mean that social sciences should borrow their theories and techniques from the natural sciences. The constitutive principles of any branch of knowledge change so rapidly that any dependent discipline is likely to continue using outmoded concepts and hypotheses. Chemists find it hard to keep up with the modern theoretical concepts of physics, and biologists still try to explain living processes in terms of outmoded physicochemical theories. Borrowing constitutive principles from another scientific discipline, furthermore, commonly leads to the habit of resorting to analogical and metaphorical explanations which contribute little to genuine understanding.

Biologists and social scientists commonly lose sight of the phenomena peculiar to life, especially to human life, because of their eagerness to use criteria that have proved useful in the physical sciences. They seem to forget that these sciences owe much of their success to the fact that they deal with phenomena free of the kind of complexities found in biological and social systems. This problem was discussed at length by Warren Weaver in his 1955 Presidential address before the American Association for the Advancement of Science:

Physical nature, first of all seems to be on the whole very *loosely coupled*. That is to say, excellently workable approximations may result from studying physical nature bit by bit, two or three variables at a time, and treating these bits as isolated. Furthermore, a large number of the broadly applicable laws are, to useful approximation, *linear*, if not directly in the relevant variables, then in nothing worse than their second time derivatives. And finally, a large fraction of physical phenomena (meteorology is sometimes an important exception) exhibit *stability*: perturbations tend to fade out, and great consequences do not result from very small causes.

These three extremely convenient characteristics of physical nature bring it about that vast ranges of phenomena can be satisfactorily handled by linear algebraic or differential equations, often involving only one or two dependent variables; they also make the handling *safe* in the sense that small errors are unlikely to propagate, go wild, and prove disastrous. Animate nature, on the other hand, presents highly complex and highly coupled systems–these are, in fact, dominant characteristics of what we call organisms. It takes a lot of variables to describe a man, or for that matter a virus; and you cannot often usefully study these variables two at a time. Animate nature also exhibits very confusing instabilities, as students of history, the stock market, or genetics are well aware. . . .

We have made small beginnings at extending the scientific method into the social sciences. Insofar as these fields can be dealt with in terms of measurable quantities, they seem to present closely intercoupled situations that can very seldom be handled with two or three variables and that often require a whole hatful—for example, W. Leontief's input-output analysis of the U.S. economy deals with some 50 variables and regrets that it does not handle more. Science has, as yet, no really good way of coping with these multivariable but nonstatistical problems, although it is possible that ultrahigh-speed computers will inspire new sorts of mathematical procedures that will be successful in cases where the effects are too numerous to handle easily but not numerous enough or of suitable character to permit statistical treatment. If we try to avoid the many-variable aspect of the social sciences by using highly simplified models of few variables, then these models are often too artificial and oversimplified to be useful. The statistical approach, on the other hand, has recently exhibited—for example, in the stochastic models for learning—new potentialities in the field of human behavior.

The theoretical concepts and methods of investigation developed for the physicochemical sciences have been highly useful of course in the life sciences, but only when applied to biological phenomena that are direct expressions of physicochemical forces. In contrast, similar methods of abstraction are less readily applicable, if at all, to behavioral and social problems.

Chemistry and physics began with the observation of gross phenomena, such as those relevant to cooking, distillation, drug use, falling bodies, and celestial movements. These sciences reached the level of mathematical abstraction only after a long period of detailed familiarity with concrete phenomena. Behavioral and social sciences probably will have to go through a phase of slowly accruing a core of concrete facts relevant to the mind and to society before they can arrive at meaningful abstract formulations of their problems. When this stage has been reached, they may re-examine their relation to natural sciences and perhaps become partly anchored on physiology, ecology, and other biological sciences.

Science and the technologies derived from it can best contribute to civilization, not through a further expansion of the mega machine but by helping in the maintenance of the ecological balance and in the development of man's potentialities. This will be made difficult by the attitudes we have inherited from the nineteenth and early twentieth centuries. We have trained our social reflexes for technological "advances," however trivial their goals and deleterious their long-range effects. Instead of conveying a teleological quality, the word "progress" now means just moving on, even though the forward motion is on a road that leads to disaster or despair. Worthwhile social goals for progress must first be formulated before planning can provide a desirable and enjoyable structure for the human effort.

Normative planning is not concerned with forecasting a future that is inevitable, but rather with "constructing" or "inventing" desirable futures—to use expressions associated with the writings of Pierre Massé and Dennis Gabor (1964). This means anticipating a desired state of affairs and acting on the present conditions to bring it about. There is a "logical" future which is the expression of natural forces and antecedent events. On the other hand, there is also a "willed" future which comes into being because man makes the effort to imagine it and to build it. H. G. Wells wrote in *A Modern Utopia*, "Will is stronger than Fact; it can mold and overcome Fact. But this world has still to discover its Will." What H. G. Wells meant by "Will" is the image of a future which is not only desired but also possible. In this light, long-range normative planning is similar to the design of utopias, except that it implies the specification of the means required to reach the utopian state.

Concern with the future used to be expressed in the form of literary exercises, or at best of purely social utopias, formulated on the basis of certain theological, political, or economic beliefs, shared by the members of the utopian group. Utopias are no longer fashionable today, partly because we lack a stable ground of generally accepted values to provide the hard foundation on which to construct viable social systems. It may be also that the eclipse of man's normative functions results from the acceptance by many scientists and sociologists of the view that the world of science and technology sets its own "arising ends". A tired resignation to the imperatives of economics and scientific technology along with the collapse of the old metaphysics may account for this acceptance. In any case the tendency during recent decades has been to limit planning to the here and now. The future is imagined not as a really new venture, but as a mere extension of the past.

To escape from this static and paralyzing view of civilized life, it will be necessary to construct multiple models of possible futures different from the present state of affairs and to imagine courses of action that would bring such futures into being. Since anticipations govern the policies of change, they paradoxically, but very effectively, become the causative agents of change. Causative anticipations differ from predictions in that the future they describe must not only be "possible" but also embody considerations of the "desirable." They imply value judgments as to what is desirable or not, good or bad, and thus inevitably give a direction to the social and scientific enterprise.

In an inspired passage in *Science and the Modern World* (1925), Whitehead suggests that the order of nature as conceived by scientific determinism has now taken the role of Fate in the Greek tragedy. The great tragedians of the modern world are the scientists "with their vision of fate, remorseless and indifferent, urging an incident to its inevitable issue . . . . This remorseless inevitableness is what pervades scientific thought. The laws of physics are the decrees of fate." Fortunately, the *applications* of science to human affairs do not have as high a degree of inevitability as do the laws of nature. Contemporary man seems to be poised between passive acceptance of scientific technology for its own sake, violent rejection of it, or conscious use of it for some ultimate concern. The social ferment which is beginning to agitate the community of scientists gives hope that man still has a chance to control his destiny by imposing a direction on the scientific endeavor and, in particular, by consciously planning the scientific technology which will shape the modern world.

Selected Bibliography

Adams, Henry. 1918. *The Education of Henry Adams*. Boston, Houghton Mifflin.

Amis, Kingsley, 1960. *New Maps of Hell: A Survey of Science Fiction*. New York, Harcourt, Brace and World.

Archer, E. James. 1968. "Can We Prepare for Famine?" *BioScience*, 18:685-690.

Aristotle. 1911. *Poetics*. Translated by S. H. Butcher. London, Macmillan. Chapter 9.

Ashby, Eric. 1960. *Technology and the Academics*. London, Macmillan.

Baldwin, James. 1963. *Notes of a Native Son*. New York, Dial Press.

Beer, Gavin de. 1965. "The Sciences Were Never at War," in *Science and Society*. Edited by N. Kaplan. Chicago, Rand McNally. Pp. 14-18.

Bell, Daniel, ed. 1967. *Toward the Year 2000: Work in Progress*. Boston, Houghton Mifflin.

Selected Bibliography

Bell, Daniel. 1964. "Twelve Modes of Prediction—A Preliminary Sorting of Approaches in the Social Sciences." *Daedalus*. 93:845-80.

Bell, Daniel. 1967. "The Year 2000—The Trajectory of an Idea." *Daedalus*. 96:639-51.

Berreman, Gerald D. 1968. "Is Anthropology Alive? Social Responsibility in Social Anthropology." *Current Anthropology*. 9: 391-96.

Blake, William. 1927. "Annotations to Sir Joshua Reynold's Discourses" in *Poetry and Prose of William Blake*. Edited by G. Keynes. London, Nonesuch Press.

Born, Max. 1968. *My Life and Views*. New York, Scribner's.

Brodine, Virginia, P. P. Gaspar, and A. J. Pallmann. 1969. "The Wind from Dugway." *Environment*, 11:2-9, 40-45.

Brooks, Harvey. 1967. "Can Science Be Planned?" in *Problems of Science Policy:* Seminar at Jouy-en-Josas on Science Policy. Paris, OECD.

Brooks, Harvey. 1968. *The Government of Science*. Cambridge, M.I.T. Press.

Brooks, Harvey. 1967. "Science and Society." *Daedalus*. 96:982-83.

Bush, Vannevar. 1967. *Science is not Enough*. New York, Morrow.

Bush, Vannevar. 1960. *Science the Endless Frontier*. Washington, D.C., National Science Foundation.

Butterfield, Herbert. 1950. *The Origins of Modern Science*. New York, Macmillan.

Calder, Ritchie, 1968. *Man and the Cosmos: The Nature of Science Today*. New York, Praeger.

Calder, Ritchie. 1964. "Common Understanding of Science." *Impact*. 14:179-195.

"Can We Keep Our Planet Habitable?" January, 1969 issue of *The Courier*, published by UNESCO. See also the new journal *Environment*.

Selected Bibliography

Cannon, W. B. 1945. *The Way of an Investigator*. New York, Norton.

Carey, W. D. 1965. "Research Development and the Federal Budget," in *Science and Society*. Edited by N. Kaplan. Chicago, Rand McNally. Pp. 409-14. See also Alfred Cote.

Chase, Stuart. 1968. *The Most Probable World*. New York, Harper and Row.

Cole, Lamont. 1968. "Can the World Be Saved?" *BioScience*, 18:679-84.

Committee on Pollution, National Academy of Sciences. 1966. *Waste Management and Control*. Washington, D.C., NAS Publication 1400.

Commoner, Barry. 1966. *Science and Survival*. New York, Viking.

Conant, J. B. 1951. *Science and Commonsense*. New Haven, Yale University Press.

Cote, Alfred J., Jr. 1968. Editorial on William D. Carey. *Industrial Research*. January, p. 17.

Crane, H. R. 1968. "Students Do Not Think Physics is 'Relevant.' What Can We Do About It?" *Amer. J. of Physics*. 12:1137-43.

Crombie, A. C. 1963. *Scientific Change*. New York, Basic Books.

Daniels, George H. 1967. "The Pure-Science Ideal and Democratic Culture." *Science*. 156:1699-1705.

Donne, John. 1919. *Donne's Sermons*. Edited by Logan P. Smith. Oxford, Clarendon Press. Pp. 100-101.

Dubos, René. 1969. "The Biosphere: A Delicate Balance Between Man and Nature." *The Courier*. January, pp. 7-16.

Dubos, René. 1968[a]. *Man, Medicine and Environment*. New York, Praeger.

Dubos, René. 1968[b]. *So Human an Animal*. New York, Scribner's.

Dubos, René. 1965. *Man Adapting*. New Haven, Yale University Press.

Dubos, René. 1961. *Dreams of Reason*. New York, Columbia University Press.

Dubos, René. 1959. *Mirage of Health*. New York, Harper.

Dubos, René. 1950. *Louis Pasteur: Free Lance of Science*. Boston, Little, Brown.

Ellul, Jacques. 1964. *The Technological Society*. New York, Knopf. See also Carl Stover.

Engstrom, Elmer. 1967. "Science, Technology and Statesmanship." *American Scientist*. 55:72-79.

Environment, Committee for Environmental Information, 438 North Skinker Boulevard, St. Louis, Missouri 63130.

Ewald, W. R., Jr., ed. 1967. *Environment for Man: The Next Fifty Years*. Bloomington, Indiana University Press.

Forbes, Robert J. and E. J. Dijksterhuis. 1963. *A History of Science and Technology*. 2 Vols. Baltimore, Penguin Books.

Foreign Policy Association. 1968. *Toward the Year 2018*. New York, Cowles.

Franklin, H. Bruce. 1967. *Future Perfect: American Science Fiction of the 19th Century*. New York, Oxford University Press.

Gabor, Dennis. 1964. *Inventing the Future*. New York, Knopf.

Galbraith, J. K. 1967. *The New Industrial State*. Boston, Houghton Mifflin.

Gillispie, Charles C. 1968. "Remarks on Social Selection as a Factor in the Progressivism of Science." *American Scientist*. 56:439-50.

Gjessing, Gutorm. 1968. "The Social Responsibility of the Social Scientist." *Current Anthropology*. 9:397-402.

Goldsmith, Maurice and Alan Mackay, eds. 1964. *Society and Science*. New York, Simon and Schuster.
Greenberg, Daniel S. 1968. *The Politics of Pure Science*. New York, New American Library.
Gundersheimer, Werner L. 1966. *The Life and Works of Louis Le Roy*. Geneva, Librairie Droz.
Hardin, Garrett. 1967. "Pop Research and The Seismic Market." *Per/Se* (Fall) p. 20.
Hardy, Godfrey Harold. 1940. *A Mathematician's Apology*. London, Cambridge University Press.
Harris, V. 1949. *All Coherence Gone*. Chicago, Chicago University Press.
Hawkes, Jacquetta. 1968. *Dawn of the Gods*. New York, Random House.
Helmholtz, Hermann von. 1962. *Popular Scientific Lectures*. New York, Dover.
Henderson, L. J. 1913. *The Fitness of the Environment*. New York, Macmillan.
Hillegas, Mark R. 1967. *The Future as Nightmare: H. G. Wells and the Anti-Utopians*. New York, Oxford University Press.
Hirsch, Jerry. 1967. "Behavior-Genetic, or 'Experiment,' Analysis: The Challenge of Science versus the Lure of Technology." *American Psychologist*, 22:118.
Hollomon, J. Herbert. 1965. "Science, Technology, and Economic Growth," in *Science and Society*. Edited by N. Kaplan. Chicago, Rand McNally. Pp. 519-27.
Holton, Gerald, ed. 1965. *Science and Culture*. Boston, Houghton Mifflin.
Hutchins, Robert. 1966. "Science, Scientists, and Politics," in *The New Technology and Human Values*. Edited by John G. Burke. Belmont, Calif. Wadsworth. Pp. 96-100.
Huxley, A. 1962. *Island*. New York, Harper and Row.

Jantsch, E. (ed.). 1969. *Perspectives of Planning.* Paris, OECD.

Jantsch, E. 1968. "Technological Forecasting for Planning and Its Institutional Implications." Proceedings of the *Symposium on National Research and Development in the 1970's.* Washington, National Security Industrial Association. Pp. 104-40.

Jouvenel, Bertrand de. 1967. *The Art of Conjecture.* Translated by N. Lary. New York, Basic Books.

Kahn, Herman and Anthony Wiener. 1968. *The Year 2000: A Framework for Speculation on The Next 33 Years.* New York, Macmillan.

Kahn, Herman and Anthony Wiener. 1967. "The Next Thirty-Three Years: A Framework for Speculation." *Daedalus,* 96:705-32.

Kaplan, N., ed. 1965. *Science and Society.* Chicago, Rand McNally.

Kash, Don E. 1968. "Research and Development at the University." *Science.* 160:1313-18.

Kennedy, J. F. 1965. Quoted in *The Scientific Endeavor.* Edited by National Academy of Sciences. New York, Rockefeller University Press.

Kerr, Clark. 1963. *The Uses of the University.* Cambridge, Harvard University Press.

King-Hele, Desmond. 1968. *The Essential Writings of Erasmus Darwin.* London, MacGibbon and Kee. Reviewed by J. G. Crowther. 1968. *Nature.* 219:655.

Koyré, Alexandre. 1968. *Metaphysics and Measurement.* London, Chapman and Hall. Reviewed by C. B. Schmitt. 1968. *Nature.* 218:1277.

Kranzberg, M. 1968. "The Disunity of Science-Technology." *American Scientist.* 56:21-34.

Kranzberg, M. and C. W. Pursell, Jr. 1968. *Technology*

in Western Civilization. New York, Oxford University Press.

Kusch, P. 1968. "The World of Science and the Scientist's World." *Bulletin of Atomic Scientists*, 24(8):38-43.

Laslett, Peter. 1965. *The World We Have Lost*. New York, Scribner's.

Le Roy, Louis. 1594. *Of the Interchangeable Course or Variety of Things in the Whole World; and the Concurrence of Arms and Learning through the First and Famousest Nations: From the Beginning of Civility and Memory of Man to this Present*. London, Charles Yetsweirt. See also Werner Gundersheimer.

Le Roy, Louis. 1575. *De la Vicissitude ou varieté des choses en l'univers, et concurrence des armes et des lettres par les premieres et plus illustres nations du monde, depuis le temps ou a commencé la civilité, & memoire humaine iusques à present*. Paris, Pierre L'Huillier.

Lindvall, Frederick C. 1968. "Science and the Social Imperatives." *American Scientist*. 56:303-11.

Lonsdale, Kathleen. 1968. "Science and the Good Life." *The Advancement of Science*. 25:1-11.

Lovejoy, Arthur O. 1936. *The Great Chain of Being*. Cambridge, Harvard University Press.

Maddox, John. 1968. "Choice and the Scientific Community," in *Criteria for Scientific Development: Public Policy and National Goals*. Edited by E. Shils. Cambridge, M.I.T. Press. Pp. 44-62.

Mandelbaum, Leonard. 1969. "Apollo: How the United States Decided to Go to the Moon." *Science*. 163:649-53.

Massé, P. 1965. *Le Plan ou L'Anti-hasard*. Paris, Gallimard.

Maurois, André. 1959. *The Life of Sir Alexander Fleming*. New York, Dutton.

Medawar, Peter. 1963. Quoted in *Man and His Future*.

Edited by G. Wolstenholme. Boston, Little, Brown. P. 382.

Merton, Robert K. 1965. "The Ambivalence of Scientists," in *Science and Society*. Edited by N. Kaplan. Chicago, Rand McNally. Pp. 113-32.

Merton, Robert K. 1961. "Singletons and Multiples in Scientific Discovery." *Proc. of the Amer. Philo. Soc.* 105:470-86.

Merton, Robert K. 1957. "Priorities in Scientific Discovery." *Amer. Soc. Rev.* 22:635-59.

Mesthene, Emmanuel G. 1968. "How Technology Will Shape the Future." *Science*. 161:135-43.

Mesthene, Emmanuel G. 1967. "The Impacts of Science on Public Policy." *Public Administration Review*. 27: 97-104.

Mesthene, Emmanuel G. 1967. "Our Threatened Planet: The Technological Plague." Review of Barry Commoner's *Science and Survival*. *Science*. 155: 441-42.

Michael, Donald N. 1968. *The Unprepared Society: Planning for a Precarious Future*. New York, Basic Books.

Nader, Claire. 1966. "The Technical Expert in a Democracy." *Bulletin of the Atomic Scientists*. 22:28-30.

National Academy of Sciences, ed. 1965. *The Scientific Endeavor*. New York, Rockefeller University Press.

Needham, Joseph. 1954. *Science and Civilisation in China*. London, Cambridge University Press.

Nef, John. 1961. "Civilization, Industrial Society, and Love." *Occasional Paper*. Santa Barbara, The Center for the Study of Democratic Institutions.

Oppenheimer, J. Robert. 1962. "On Science and Culture." *Encounter*, 19:1-8. See also I. I. Rabi, *Oppenheimer*, 1969.

Ortega y Gasset, José. 1932. *The Revolt of the Masses*. New York, Norton.

Ozbekhan, H. 1969. "Toward a General Theory of Planning" in *Perspectives of Planning,* E. Jantsch (ed.). Paris, OECD.

Ozbekhan, H. 1968. *The Idea of a 'Look-Out' Institution.* Paris, Futuribles.

Pohl, F. 1957. *The Case Against Tomorrow.* New York, Ballantine.

Polanyi, Michael. 1968. "The Republic of Science: Its Political and Economic Theory," and "The Growth of Science in Society," in *Criteria for Scientific Development: Public Policy and National Goals.* Edited by E. Shils. Cambridge, M.I.T. Press. Pp. 1-20 and 187-99.

Predmore, R. L. 1968. "What Role for the Humanist in These Troubled Times?" *BioScience.* 18:691-93.

Price, Don. 1969. "Purists and Politicians." *Science.* 163: 25-31.

Price, Don. 1965. *The Scientific Estate.* Cambridge, Belknap Press.

Rabi, I. I., Robert Serber, Victor F. Weisskopf, Abraham Pais and Glenn T. Seaborg. 1969. *Oppenheimer.* New York, Scribner's.

Rabi, I. I. 1965. Quoted in *The Impact of Science on Technology.* Edited by A. W. Warner, Dean Morse, and A. S. Eichner. New York, Columbia University Press. P. 13.

Remer, Theodore G., ed. 1965. *Serendipity and the Three Princes.* Translated by A. G. and T. L. Borselli. Oklahoma City, University of Oklahoma Press.

Reston, James. 1967. "Washington: The New Pessimism." *The New York Times.* April 21, 1967. P. 38. Also the editorial "Voices of Doubt," in *The Wall Street Journal.* April 26, 1967. P. 16.

Restoring the Quality of our Environment. 1965. Report of the Environmental Pollution Panel, President's Science Advisory Committee, Washington, The White House.

Roberts, Walter Orr. 1967. "Science, A Wellspring of our Discontent." *American Scientist*. 55:3-14.

Schon, Donald A. 1967. *Technology and Change*. New York, Delacorte Press.

Seitz, Frederick. 1966. Dedication Address for the Kline Biology Tower at Yale University. October 28.

Shannon, James. 1966. "Biomedical Sciences—Present Status and Problems," in *Science, Government and the Universities*. Introduced by F. Seitz. Seattle, University of Washington Press. Pp. 61-70.

Shils, Edward, ed. 1968. *Criteria for Scientific Development: Public Policy and National Goals*. Cambridge, M.I.T. Press.

Skinner, B. F. 1948. *Walden Two*. New York, Macmillan.

Smith, Cyril Stanley. 1968. "Matter versus Materials: A Historical View." *Science*. 162:637-44.

Snow, C. P. 1959. *The Two Cultures and the Scientific Revolution*. New York, Cambridge University Press.

Sprat, Thomas. 1667. *History of the Royal Society of London*.

Stevenson, Adlai. 1965. Speech given before the Economic and Social Council. Geneva. July 9.

Steward, Julian H. 1968. "Scientific Responsibility in Modern Life." *Science* 159:147-48.

Stover, Carl, ed. 1963. *The Technological Order*. Detroit, Wayne State University Press.

Stratton, Julius A. 1966. "The M.I.T. 1964 Commencement Address," in *The New Technology and Human Values*. Edited by John G. Burke. Belmont, Calif., Wadsworth. Pp. 92-96.

Sykes, Gerald. 1967. *The Cool Millennium*. Englewood Cliffs, Prentice-Hall.

The Task Force of Environmental Health and Related Problems. 1967. A *Strategy for a Livable Environment*. Washington, U. S. Dept. of Health, Education and Welfare. June.

Taton, R. 1962. *Reason and Chance in Scientific Discovery*. New York, Science Editions.

Tolstoy, Leo. 1898. "The Superstitions of Science." *The Arena*. 20:52-50. Reprinted in *The New Technology and Human Values*. Edited by John G. Burke. Belmont, Calif., Wadsworth. Pp. 24-30.

Toulmin, Stephen. 1968. "The Complexity of Scientific Choice: A Stock-taking," and "The Complexity of Scientific Choice II: Culture, Overheads or Tertiary Industry?" in *Criteria for Scientific Development: Public Policy and National Goals*. Edited by E. Shils. Cambridge, M.I.T. Press. Pp. 63-79 and 119-33.

Toulmin, Stephen. 1961. *Foresight and Understanding*. Bloomington, Indiana University Press.

Townes, Charles H. 1968. "Quantum Electronics and Surprise in Development of Technology." *Science*. 159:699-703.

Toynbee, Arnold J. 1968. "Science in Human Affairs: An Historian's View." *Occasional Paper, The Institute for the Study of Science in Human Affairs*. New York, Columbia University.

Udall, Stewart. 1966. "Can America Outgrow its Growth Myth?" Address before the Long Island Conference on Natural Beauty. Hofstra University.

Unamuno, Miguel de. 1954. *Tragic Sense of Life*. Translated by J. E. Crawford Flitch. New York, Dover.

UNESCO. 1968. International Conference on "Man and the Biosphere." Paris, September.

U.S. Congress, House Subcommittee on Science, Research and Development of the Committee on Science and

Astronautics. 1966. *Inquiries, Legislation, Policy Studies Re: Science and Technology*, 2nd Progress Report, 89th Congress, 2nd Session. Washington, U.S. Government Printing Office.

Vucinich, Alexander. 1968. "Science and Mortality: A Soviet Dilemma." *Science*. 159:1208-12.

Walker, Patrick Gordon. 1966. "The Origins of the Machine Age." *History Today*. 16:591-600.

Weaver, Warren. 1955. "Science and People." *Science*. 122:1255-59.

Weinberg, Alvin M. 1968 "Criteria for Scientific Choice," "Criteria for Scientific Choice II: The Two Cultures," and "Scientific Choice and Biomedical Science," in *Criteria for Scientific Development: Public Policy and National Goals*. Edited by E. Shils. Cambridge, M.I.T. Press. Pp. 21-33, 80-91, and 107-118.

Weinberg, Alvin M. 1967. *Reflections on Big Science*. Cambridge, M.I.T. Press.

Wheeler, Harvey. 1969. "Bringing Science Under Law." *The Center Magazine*. 2:59-67.

Wheeler, Harvey. 1968. "The Constitutionalization of Science." *Dialogue*. Santa Barbara, The Center for the Study of Democratic Institutions. October 2.

White, Lynn, Jr. 1967. "The Historical Basis of Our Ecological Crisis." *Science*. 155:1203-07.

Whitehead, Alfred North. 1925. *Science and the Modern World*. New York, Macmillan.

Wiener, Norbert. 1966. "The Monkey's Paw" in *The New Technology and Human Values*. Edited by J. Burke. Belmont, Wadsworth. Pp. 130-34.

Wiener, Norbert. 1950. *The Human Use of Human Beings: Cybernetics and Society*. Garden City, Doubleday.

Wilkinson, John. 1967. "Futuribles: Innovation vs. Stability." *Center Diary:* 17, March-April.

Wilkinson, John. 1966. *Introduction to Technology and Human Values*. Santa Barbara, The Center for the Study of Democratic Institutions.

Wirt, John. 1968. "Students, Technology and New Priorities." *Stanford Alumni Almanac*, December.

Wolstenholme, G., ed. 1963. *Man and His Future*. Boston, Little, Brown.

Yefremov, Ivan. 1957. *Andromeda*. Translated by George Hanna. Quoted in Mark Hillegas. 1967. *The Future as Nightmare*. New York, Oxford University Press.

Ziman, J. M. 1968. *Public Knowledge*. London, Cambridge University Press.

Index

Index

Index

Index